Data Communications for Microcomputers

SO-BZV-645

With Practical Applications and Experiments

Elizabeth A. Nichols *Digital Analysis Corporation*

Joseph C. Nichols *Digital Analysis Corporation*

Keith R. Musson *ConTel Information Systems*

McGraw-Hill Book Company

New York St. Louis San Francisco Auckland Bogotá
Hamburg Johannesburg London Madrid Mexico
Montreal New Delhi Panama Paris São Paulo
Singapore Sydney Tokyo Toronto

This book is dedicated to:
Richard K. Agnew, Jr.
Peter K. Agnew
Cynthia N. Kettyle
Rosemary P. Musson

Library of Congress Cataloging in Publication Data

Nichols, Elizabeth Agnew.
  Data communications for microcomputers.

  Includes index.
  1. Data transmission systems.   2. Microcomputers.
I. Nichols, Joseph C., 1943–   II. Musson,
Keith R.   III. Title.
TK5105.N523   001.64′4   81-17141
ISBN 0-07-046480-4   AACR2

    567890 DODO 89876543

ISBN 0-07-046480-4

The editors for this book were Barry Richman, Stephen Guty, Susan Thomas;
the designer was Naomi Auerbach; and the production supervisor
was Paul A. Malchow. It was set in Aster by University Graphics, Inc.

Printed and bound by R. R. Donnelley & Sons Company.

Contents

4 *Asynchronous-Serial-Related Conventions and Experiments* 145

5 *Modems* 221

Preface

It is the intention of this book to address data communications between equipment most often found in microcomputer systems environments. In particular, local data communications between microcomputers, keyboard/display units (or terminals), and printers is addressed as well as remote communications between these entities via the public-switched telephone network.

This book addresses data communications in microcomputer-based systems at several levels. The first level pertains to *practical* information. How do you make a cable? What parts must you purchase? How do you write software to perform input/output? What are some useful utilities and how can they be implemented? The second level is a tutorial on the concepts and principles of elementary data communications. The underlying principles behind the practical techniques are presented so that readers will have the information to build upon what has been presented here. The third level is a reference tool. Our objective is to provide between the covers of one book enough charts, tables, standards information, timely references, and so on to support most microcomputer-based data-communications projects. Therefore, we have included a great deal of reference material. The text of the book supports this reference material by providing examples of how to use it.

The mechanism for addressing the practical level of data communications must be lots of examples and experiments. Many of the experiments, such as those that address cable configurations, cable assembly, or the development of accessory equipment for troubleshooting data-communications problems, are independent of particular systems and are applicable to virtually all microcomputer-based systems.

On the other hand, some experiments, by necessity, are explained in a system-dependent fashion. Those experiments and software modules that are system-dependent are based on one of two classes of microcomputer systems: TRS-80* BASIC systems and Z80 CP/M† (control program/microcomputer) systems. Several of the experiments make use of particular hardware and software in one of the above classes. Most of these experiments and much of the software that is presented can be converted relatively easily to execute on systems other than CP/M or TRS-80 systems.

Thus, via a combination of material ranging from excerpts of international standards publications to parts lists from your local computer store, we have attempted to provide one approach to data communications for a microcomputer environment.

Joseph C. Nichols
Elizabeth A. Nichols
Keith R. Musson

---

*TRS-80 is a trademark of Tandy Corporation.
†CP/M® is a registered trademark of Digital Research, Inc.

Acknowledgments

The authors wish to express their thanks to Barry Richman of McGraw-Hill for his help in the preparation of this book, the Electronic Industries Association for their permission to reprint EIA standards material, and SGS-ATES Componenti Elettronici SpA for their permission to reprint selected technical material from their publications.

1 *Background and Perspective*

This book is concerned with microcomputers and data communication. Why microcomputers as opposed to other classifications of digital computers? The essential principles for digital communications remain the same across all types of digital computer systems, but there are several properties of microcomputer systems that make their problems unique. One important characteristic relates to who is responsible for initially configuring and maintaining a microcomputer system.

Consider the case of Mr. I. C. Mos. I. C. is a traveling salesman for the Solar Widget Corporation who has had experience accessing the company computer for several years. He has used it extensively both in the office and on the road to perform routine bookkeeping tasks associated with his sales and customer support.

I. C. has thought vaguely about the many potential advantages of owning his own microcomputer system. He would like to configure a system that would provide him access to the company computer from his home. Moreover, he would like to be able to query the UPI data bases for up-to-the-minute sports results and the evening stock exchange information. I. C. is aware that there are several other similar time-sharing systems that he could access if only he had the means. Since a microcomputer system possesses significantly more local resources than a dumb terminal, I. C. is quite excited about the prospect of exchanging data with and utilizing the power of remote systems through his own private facility in his home. In addition, I. C. would like a system that can support word processing of some form and provide access to an electronic mail system.

A market survey is conducted. I. C. concludes that he will obtain the hardware components for his system from two separate vendors: TCI (The Computer Igloo), located in his home town, and FEMCO (Fast Eddie's Micros). I. C. has also decided to obtain additional applications software from The 4S Company (Suzzie's System Software and Storm Door Company). Although these three companies spent a total of 12 ms selecting their company names, their systems prove to be superb. The potential problem turns out to be one of integrating the three vendors' "subsystems" into one functioning unit.

However, I. C. decides that he can handle the necessary systems integration to configure his optimal system based upon the three vendors. I. C. knew from his mainframe systems experience that input/output operations comprise the most installation-dependent aspect of hardware and/or software systems, but the ensuing months of microcomputer system integration gave him a new perspective on this fact. His first experience began when he had to come up with a cable to attach his keyboard/display unit to the computer. When his first attempt did not produce anything even remotely functional, he called the customer service department at FEMCO, where he purchased his keyboard/display unit. The service technician listened carefully to the symptoms and then dazed I. C. with a sequence of questions. It all happened very fast, but I. C. remembers that the technician asked him something about his cable conforming to the RS-232-C standard, something about putting his scope on the Receive pin of the USART, and something about using the XON convention. As far as I. C. Mos was concerned, RS-232-C was a license plate number, USART was a federal agency that preserved rare paintings, and XON was an international oil company that was getting into the information-processing business.

His second experience came when he started planning how to access remote systems. He discovered that getting his brand X microcomputer to communicate with a brand Y time-sharing system was not absolutely trivial. His functional requirements were simple, and he was willing to settle for any quick and dirty utility that would consistently transfer disk files between his system and the remote host. Unfortunately, detailed information on the type of software required to accomplish this was difficult to find, and all the microcomputer consultants that I. C. knew were charging over $35 per hour for their services.

Mr. I. C. Mos's experiences are not unique. The initial configuration and daily operation of microcomputer systems are frequently the sole responsibility of the end user. The low cost of the system may not justify the kind of comprehensive vendor-supplied installation support that is available for minicomputers and mainframes. As mentioned above, I/O functions are the most installation-specific aspect of computer system implementation.

THE MICROELECTRONIC REVOLUTION

Interesting new words are appearing in speeches, news magazines, technical journals, and the public vocabulary. The *plugged-in home*, the *paperless office*, the *information utility*, and *telecommuting* are some of the words and phrases that are being used to describe the future, and, indeed, the present. The incontrovertible evidence is that data communications and computer networking are finding their way into the office, the laboratory, the classroom, and the home. This trend is, in large measure, due to the phenomenon known as *microelectronics*, which has accelerated the proliferation of consumer-oriented devices, low-end computer components, and inexpensive communications equipment. An increasing proportion of these products have the capability to exchange information with other similar devices and with large centralized facilities. The data-transmission mechanisms include the use of the public-switched telephone network, the use of cable TV transmission facilities, and the use of broadcast signals such as AM, FM, UHF, and VHF. The applications to which these intercommunicating systems are put include information storage and retrieval, word and document processing, electronic mail, games and entertainment, remote control of devices for security and energy savings, and many, many more.

The population of individuals who will be affected by this invasion of computer and communications technology into almost all routine daily activities can be characterized as members of at least one of the following classes: creators, knowledgeable users, and blissfully ignorant consumers. The creators are practicing professional engineers, computer scientists, and data-communications specialists with detailed knowledge about specific aspects of the technology and general knowledge about most of it. The blissfully ignorant consumers are end users who do not care about how it works or how to enhance it, but rather have strong feelings about when it fails. This book is primarily targeted at individuals who wish to become members of the knowledgeable user population. The present state of the art and newness of the technology dictate that the ignorant user will not remain blissful for long. Thus, our purpose in selecting and presenting the material contained herein is to help those current and potential users of data communications in a microcomputer system environment become more knowledgeable, so that they can capitalize on the powerful capabilities that are available if they just know where to look.

In this chapter we will discuss the idea of the information utility and one of the types of common carrier services, the public packet-switched networks that help to provide public access to possibly remote information distribution and processing centers. The desire to interconnect and integrate heterogeneous systems imposes strict requirements for interface standards at several different levels of communication. Many organizations exist that define and review standards relevant to communications and computer networking. In the last section of this chapter, we present a list of the major standards organizations with such information as how to obtain their literature and what specific standards they promulgate which are related to data communications in a microcomputer environment.

OBJECTIVES

When you have completed this chapter, you will be able to do the following:

- Be aware of the emergence of products and services that implement the idea of information as a utility.
- Be able to identify and briefly characterize the services offered by information utilities.
- Define the concept of a public packet-switched network and identify several such existing networks.
- Identify several standards organizations and their relevant work in the area of data communications for microcomputers.
- Be familiar with some of the terminology and concepts associated with the International Standards Organization's open systems interconnection (OSI) model for networking with heterogeneous equipment.

INFORMATION AS A UTILITY

A *utility* may be defined as a provider of a necessary and valuable service. Classically, utilities have provided such services as the distribution of electricity, gas, and water to the place where it is to be consumed. A new commodity to be added to this list is information. The information utility, for a price, performs the service of delivering information that is timely, organized, and tailored to the user's unique needs. The information is delivered directly to the user, almost immediately after it is requested, no matter where the user is physically located.

An additional important characteristic of the information utility is that it is essentially an "information broker." The information utility itself merely supplies the vehicle for storing and distributing the information, while it purchases the actual data from other sources. In this manner, the utility need not worry about the data-collection problem and the information source need not worry about the dissemination problem. Both parties do what they do best; it's a truly synergistic relationship.

Currently, there are two major categories of information utility: the time-sharing-based service and the Viewdata/Teletext type of service. The time-sharing-based service is an outgrowth of the computer time-sharing industry that has existed since the late 1960s. In the United States, the two principal representatives of this class are The Source and Micronet.

The Source and Micronet

The Source was established in 1979 by Telecommunications Corporation of America (TCA) in McLean, Virginia. The resources of large time-sharing services may be significantly underutilized outside normal business hours. This same underutilization of resources outside normal business hours applies to common carriers such as the public packet-switched networks, Tymnet and GTE Telenet, which can provide relatively inexpensive access to these time-sharing services. The common carriers and their services are discussed in more detail in the following section of this chapter.

The Source provides access to large data bases through the facilities of the Dialcom time-sharing system in Silver Spring, Maryland. Included among the original Source information providers were United Press International (UPI), *The New York Times*, and the major stock and commodities exchanges. Thus, the three necessary ingredients for an information utility were in place—vast amounts of information, a computer system through which this information could be accessed, and a relatively low cost distribution network accessible by virtually every home and business in the United States. Most potential Source users could be connected to the value-added GTE Telenet and Tymnet networks via a local telephone call, and thence to Dialcom in Maryland.

In order to become a Source subscriber, an individual must have a low-speed modem and an ASCII keyboard/display terminal. Modems, the ASCII (American Standard Code for Information Interchange) character set, and the cables and connectors that are required for connection to The Source are discussed in great detail in subsequent chapters of this book. The cost of such equipment is approximately $1000. The hourly rate quoted by The Source in early 1981 for system access during the 6 p.m. to 7 a.m. time period for weekdays and at any time during weekends ranges from $2.75 per connect hour to $4.25 per connect hour. This compares to a rate of $15 per connect hour charged to users during prime time normal business hours. The charge for data storage was quoted in early 1981 at $0.016 per day for a block of 2048 characters. This translates to about $0.50 per kilobyte block per month.

The Source offers services in addition to data-base access. Almost all of the standard time-sharing services are also available. These include compilers for program development, data-base management utilities, text editors and document processors, electronic mail, games and specialized applications programs for functions such as accounting, and statistical analysis. In all, over 2000 programs and utilities are available from The Source.

Micronet is another early manifestation of the information utility. A subsidiary of the nationwide time-sharing company Compuserv, Micronet uses its parent corporation's network to provide user access to its resources. In terms of services and connect charges, Micronet and The Source compete directly in the home and small business user markets.

Viewdata and Teletext

The second major class of information utilities are called *Viewdata* and *Teletext* services. Both use a color television screen to display data

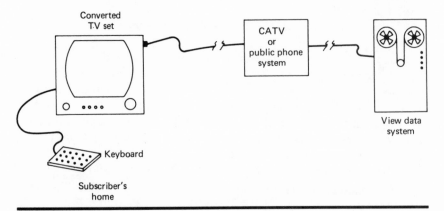

Fig. 1-1 Viewdata approach.

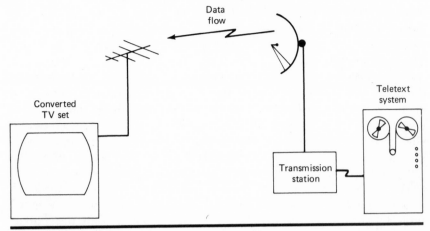

Fig. 1-2   Teletext approach.

arranged into discrete pages. Both Viewdata and Teletext were first implemented in the United Kingdom—Viewdata by the British Post Office as a simple distribution mechanism that could later be enhanced, and Teletext by the British Broadcasting Corporation as a system to broadcast captions for the deaf.

There are two significant differences between types of systems like Viewdata and Teletext: the data-distribution technology employed and interactivity. Viewdata systems use the telephone network, while Teletext systems use television broadcast media to transfer data from a central facility to the user's television set. Viewdata supports two-way communications, while Teletext does not.

Since Viewdata and Teletext systems are not yet widely available in the United States and do not employ general-purpose programmable microcomputers in the home or office, they are not directly applicable to the purpose of this book. We mention them because they are playing an increasingly important role in the developing and marketing of information.

VALUE-ADDED COMMUNICATIONS NETWORKS

The ideas of plugged-in homes, paperless offices, and information utilities are predicated upon readily available, inexpensive data communications. The oldest and most universally available communications resource is the public-switched telephone network. Virtually every home and office in the United States has a telephone, and, therefore, the capability to communicate across this network with any other similarly equipped establishment.

In this section we will outline some of the common carrier services that are available now or are currently in the development stage. Some of these common carrier services are clearly targeted at the large corporate user. However, even though a service is initially targeted only at large corporate users, eventually some entrepreneur may subscribe to the service as a "large user" and resell elements of the service to smaller users. Eventually this process may filter down to Mr. I. C. Mos.

In the early 1970s the Department of Defense Advanced Research Projects Agency announced plans for building a computer communications network called the *Arpanet*. The Arpanet used the public-switched telephone network to interconnect many different computers at locations across the United States. The Arpanet also used a technique called packet switching to route messages between communicating nodes in the network. *Packet switching* is a technique in which messages are divided into fixed-length segments called *packets*. The packets that comprise a particular message are then routed individually through the network to their ultimate destination. The packetizing technique creates a uniform increment of network workload and facilitates the buffering and routing of messages at the network switching nodes. The result is that packet-switching networks can achieve high throughput in a very cost effective manner. Detailed performance data collected on the Arpanet confirmed that packet-switching technology could be an effective, commercially viable solution to interconnecting large numbers of heterogeneous systems via the public-switched telephone network.

Therefore, by the mid-1970s, there was strong evidence that the telephone network, coupled with specialized hardware and software to implement packet switching, could be integrated and marketed to a growing population of users in need of (relatively) inexpensive computer communications capacity. The resulting product is called the *value-added network*. In particular, a value-added network supplier leases telephone circuits from the telephone companies and provides the packet-switching hardware and software for multiple companies and users to share in a cost-effective manner.

Two major commercial value-added network companies are GTE Telenet and Tymnet. Each of these companies implements its own packet-switched telephone-system-based networks. The "value-added" adjective refers to the packet-switching capability plus additional network services such as redundant equipment, alternative routing if a particular link in the network fails, network resource management that effectively cuts communications costs to about one-third of those associated with private leased telephone circuits, and transmission-error control that reduces bit error rates to levels much, much lower than those normally occurring over a standard telephone line. Additional services, such as electronic mail, are also offered by value-added carriers.

By the late 1970s, several new network services were announced by leading office products and communications companies—all designed to take advantage of the latest technologies, such as microwave and satellite communications. American Telephone and Telegraph Company (AT&T) and Satellite Business Systems (SBS) are two such companies.

The question arises: What does this have to do with data communications for microcomputers? The answer is that home users and small microsystem users are already tied in to the resources and services provided by value-added network companies via information utilities such as The Source. In particular, The Source utilizes GTE Telenet and Tymnet to connect users located all over the United States to the large computer resources of a time-sharing service in Silver Spring, Maryland. The clear trend is that new services that are at first too expensive for small companies and individual users do eventually find their way into the small consumer market—either as a result of cost reductions or as a result of the effective uniting, by intermediate entrepeneurs, of large populations of individuals as one larger customer.

As the information-utility concept begins to invade more offices and households, there are associated growing concerns about standardization. Similar problems have occurred in other industries, such as the videotape recorder/player market. To point out an example of the benefits of standardization, notice that it matters not whether you buy a television set from vendor A or vendor B. You will still be able to watch Monday Night Football. It is extremely important for a growing industry, such as the data-communications and information-services industries, to define workable standards and remain consistent with them. If users are constantly faced with choices among products that present compatibility problems, the effects will be negative for everyone concerned. Therefore, a number of standards organizations exist whose purpose is to design and develop workable standards plus provide public forums for the discussion and resolution of standards issues. The next section provides a brief description of several standards organizations that have significance with respect to data-communications and microcomputer systems.

STANDARDS ORGANIZATIONS

There are two American organizations and two international organizations that have significant impact upon users of microcomputers in data-communications applications. The following paragraphs provide brief descriptions and identify particularly significant standards, relevant to our context, that are promulgated by each.

EIA

The Electronic Industries Association (EIA) is an organization of U.S. manufacturers in the electronics industry. EIA standards work in the data-communications area is performed by Technical Committee TR30. The EIA publishes standards in its RS-series, with additional applications notes and supplementary materials appearing in the *Industrial Electronics Bulletin.*

The EIA standard that is most applicable to microcomputer systems is RS-232-C, which governs the electrical characteristics for connecting terminals to modems, terminals to computers, printers to computers, and other similar pairs of computer components. Recent EIA standards are RS-449, RS-422-A, and RS-423-A. All these standards are discussed in Chapter 3.

EIA publications can be ordered from the EIA national headquarters at the following address:

Electronic Industries Association
2001 Eye Street, N.W.
Washington, DC 20006

IEEE

The Institute of Electrical and Electronics Engineers (IEEE) is a professional organization of individuals interested in electrical and electronics engineering. IEEE standards that may be applicable to microcomputer users include the IEEE S-100 bus standard and the P896 Subcommittee Proposal ("Advanced Microcomputer System Backplane Bus"). The IEEE Computer Society's Microprocessor Standards Committee may be contacted at the following address:

IEEE Computer Society
10662 Los Vaqueros Circle
Los Alamitos, CA 90720
(714) 821-8380

CCITT

The Consultative Committee in International Telegraphy and Telephony (CCITT) is a committee within the International Telecommunications Union (ITU), which is an agency of the United Nations. Two study groups within the CCITT work on standards related to data communications. CCITT Study Group VII develops standards for data communications over public data networks, such as the value-added networks that we discussed earlier in this chapter. Its work is published in ISO X-series recommendations. The second study group, Study Group XVII, develops standards related to communications over telephone facilities, and its reports are published in ISO V-series documents.

The CCITT recommendations that are most applicable to data communications for microcomputers are V.28 (essentially compatible with RS-232-C), V.10 (compatible with EIA RS-423-A), V.11 (compatible with EIA RS-422-A), and X.21 (essentially compatible with EIA RS-449).

ISO and ANSI

The International Standards Organization (ISO) is a voluntary worldwide federation composed of the principal standardization institute from each member nation. The United States' ISO representative is the American National Standards Institute (ANSI).

The objective of the ISO is to draft and gain consensus for international standards. The work for developing international standards is performed by ISO technical committees. ISO work in the data-communications areas is concentrated in two subcommittees of Technical Committee 97, which studies computers and information processing. The first subcommittee, ISO/TC97/SC6, is responsible for development of standards in the area of data communications. The second subcommittee, ISO/TC97/SC16, is developing a model for interconnecting heterogeneous systems called the *open systems interconnection (OSI)* model.

There are ISO standards that are compatible with the CCITT and EIA standards mentioned above. For example, ISO 2110 is essentially compatible with EIA RS-232-C and RS-366-A (which covers automatic calling units), and ISO 4902 is compatible with EIA RS-449.

The committee of ANSI that is concerned with data communications is called the Committee on Computers and Information Processing X3. The ANSI standard that is most applicable to microcomputer systems users is its 7-bit character encoding standard called the American Standard Code for Information Interchange (ASCII). Chapter 4 contains a section that discusses this standard in detail.

ANSI publications can be purchased by writing to one of the following addresses:

Computer and Business Equipment
  Manufacturers Association
  1828 L Street N.W.
  Suite 1200
  Washington, DC 20036

American National Standards
  Institute
  1430 Broadway
  New York, NY 10018

Note that the Computer and Business Equipment Manufacturers Association (CBEMA) is an organization that serves as secretariat to ANSI.

THE ISO REFERENCE MODEL FOR OPEN SYSTEMS INTERCONNECTION

As we mentioned above, the ISO is currently developing a standard to define a universal reference network architecture for interconnecting heterogeneous computer systems. The ISO Subcommittee 16 (SC16), which was chartered in 1977, has proposed an open systems interconnection reference model for an architecture designed to form the basis for defining future distributed systems standards. The OSI reference model is primarily applicable to large computer networks, but some of its concepts are germane to much of the material presented in this book. Our purpose in presenting a *very brief* discussion of it here is to introduce some basic concepts of networking and computer protocols and to define some terms that are important to understanding much of the literature on these topics.

Communications Protocols

In the late 1970s, the general populace became aware of the jargon used by truckers using the CB (citizens' band) radio. Such phrases as "bear in the air" and "keep the clean side up and the dirty side down" might mean nothing to the untrained ear, but to the truckers they are part of a highly structured language with well-defined rules and meanings. A typical transmission from one trucker to another begins with the first trucker giving his own name ("handle") and possibly that of the trucker that he wishes to communicate with. Correspondents give and receive confirmation of information received and formally sign off the channel when they are finished.

This type of conversation is considerably different from a normal face-to-face discussion between two truckers. The reasons for this are several. First, many users are sharing the communications resource. Thus, users must identify themselves and those to whom they are directing their remarks. Moreover, since the medium can support only limited communications traffic, messages need to be precise, or coded, and short. Since externally induced noise and interference commonly impair the clarity of a message, confirmation and, if necessary, repetition of messages are standard procedure.

A collection of such rules, procedures, conventions, and language is called a *protocol*. Before the Arpanet project in the early 1970s, the term *protocol* was used mainly in diplomatic applications in which the maintenance of appropriate decorum, customs of behavior, and social practices among different cultures was highly important. In the Arpanet project, the heterogeneous set of geographically distributed computers that had to communicate with each other had the same problem that the diplomats (and truckers) do, namely that of multiple users with a need to send information through a noisy, hostile environment over a channel with limited capacity (resulting in possibly long delays). Thus, the designers of the Arpanet saw the need for structuring and controlling communications between computers and applied the term *protocol* to this concept.

Protocols and Network Functions

In the diplomatic sense of the word *protocol*, communications between countries are carried out on multiple layers, with each layer using its own specific protocol. In computer communications protocols, the same functional layering turns out to be very useful. The ISO OSI reference model identifies seven functional layers. These seven layers can be briefly described as follows.

Physical Layer

The physical layer represents conventions directly applicable to the physical media that interconnect the two correspondents. In particular, this layer must provide the mechanical, electrical, and functional mechanisms for the initial establishment, maintenance, and ultimate release of a physical channel linking the source and destination. In this book, we will discuss several physical-layer communications protocols: EIA RS-232-C, RS-449, RS-422-A, and RS-423-A. All these protocols are covered in Chapter 3.

Data-Link Layer

The data-link layer is responsible for handling the transfer of units of information from one node to a second node across one communications link. An important link layer protocol for microcomputer systems is the asynchronous serial data-link protocol. This topic is discussed in Chapter 2.

Network Layer

The network layer controls the switching and routing of information across the network. Thus, the network layer establishes the physical and logical connections required to transfer the data from its source to its destination(s).

Transport Layer

The transport layer is responsible for ensuring high-quality network service. One important function performed by the transport layer is controlling end-to-end data integrity. The overall purpose of the transport layer is to act as an interface between the session and network layers to ensure that the network layer provides high-quality service to the session and higher-level functions of the network.

Session, Presentation, and Application Layers

The session, presentation, and application layers are all concerned with high-level network functions. The session layer is responsible for coordinating the interaction between the processes associated with the two communicating applications. The presentation layer performs any necessary formatting and/or code conversion to make the transmitted information recognizable to the destination. The application layer is the layer that communicates directly with the user's application program or process.

For each of the above functional layers, there is an associated peer protocol that governs communication between functions in corresponding layers. The two layers applicable to general-purpose microcomputer systems that are addressed in detail in this book are the physical layer and the link layer.

2 *Data Transfers*

When computers were first built in the late 1940s, the heavy design emphasis was on *compute power*. The earliest computers were designed to solve problems that were primarily scientific in nature and, as such, were taxing the mathematical and logical capabilities, rather than the I/O resources, of the machines. Today, in selected applications, such as weather prediction, nuclear research, and economic modeling, it is possible to see several million dollars worth of super compute power (e.g., the Cray computers or some of the larger Control Data Corporation computers) interfaced to a small $2000 30 character per second printer. This type of configuration embodies the classical compute-bound application: the computer arithmetic and logic unit "crunches" numbers upon numbers, periodically blinking some lights on its console, and maybe clattering out a few "answers" on a rickety old teleprinter.

The compute-bound application, although certainly interesting and significant, is not consistent with the vast majority of applications with which general-purpose microcomputer systems are associated now. It is interesting to note that the compute power in most microcomputer systems totally resides on a single silicon chip. In this book, our examples address the Z80 microprocessor that was designed and first produced by Zilog Corporation. A rather elementary principle that every computer science student has heard is that a computer application can be subdivided into three major subfunctions: input, process, and output. If we consider only the hardware requirements associated with supporting these three subfunctions, the microprocessor CPU (central processing unit) integrated circuit essentially covers the process function. What about input and output? As computer manufacturers discovered, to an ever-increasing extent as digital machinery found its way into government, business, and education, I/O poses some very significant problems. In fact, in the last few years, the computer industry has undergone a total reversal in a fundamental systems-design axiom. In the early days, the logic was expensive, and the I/O was comparatively cheap. Now, quite the opposite holds true: the logic is cheap and the communications is expensive. Consider your microcomputer system. Where have you sunk the really big bucks? The letter-quality printer? The floppy disk? The CRT terminal? What did your CPU cost? A Z80-CPU chip today costs about $10. A processor printed circuit card costs $200 to $400.

So, this book is about the expensive part of microcomputer systems development and applications. In particular, this chapter addresses the general topic area of data transfers. The objectives of this chapter may be summarized as follows.

OBJECTIVES

When you have completed this chapter, you will be able to:

- Characterize and discuss microcomputer data transfers in terms of serial versus parallel transfers and in terms of how the transfers are controlled.
- Identify and define a hierarchy of data-transfer control techniques that incorporates:
  Program control
  - Polled I/O
  - Interrupt-driven I/O
  Direct memory access (DMA)
- Discuss DMA in terms of its benefits, disadvantages, and appropriate applications.
- Discuss polled I/O and interrupt-driven I/O in terms of their relative strengths and weaknesses.
- Describe asynchronous serial protocol in terms of each type of bit transmitted: start, data, parity, and stop.

MICROPROCESSOR INTERFACING

As we mentioned above, the computational and logical capabilities of a microcomputer reside on a single integrated circuit, sitting on a printed circuit card that may be plugged into the backplane of the computer cabinet. Therefore, the problem of performing data transfers with this small component essentially reduces to a concept known as *microprocessor interfacing*. The following is a formal definition of this term:

MICROPROCESSOR INTERFACING: The integration of a microprocessor with memory, I/O devices, and other external components to function in a compatible, coordinated fashion.

The goal is to effectively communicate data to the microprocessor, and then, after the CPU has appropriately processed the data, provide a mechanism by which the CPU may communicate the results. The essential components in this process are illustrated in Figure 2-1.

Any data transfer must involve a source, a destination, the data itself, and a mechanism for control of the transfer. This can be boiled down to three key facts: *where, what,* and *when.* The *where* gives the addresses of the data source and destination. The *what* comprises the data itself. The *when* covers the control mechanism for synchronizing the events that must occur to achieve the transfer.

The data source and destination may be any pair from the following set of three possibilities, in any combination: CPU, memory, peripheral. Therefore, possible source-destination pairs are two CPUs, a CPU and memory, a CPU and a peripheral, two peripherals, memory and a

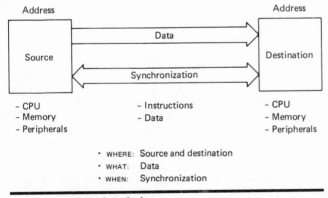

Fig. 2-1 Computer interfacing.

peripheral, etc. The data that is transferred may consist of instruction bytes or actual data.

PARALLEL AND SERIAL DATA TRANSFERS

In the above discussion of microprocessor interfacing, we cited three major elements, where, what, and when. In this section we will discuss the important concept of the *mode* of transmission. In the computer world, there are two data-transfer modes: *serial transfers* and *parallel transfers.*

The data that comprise the *what* of a transfer are typically quantized as *groups* of *bits.* Figure 2-2 illustrates several bit groupings that are most common for microcomputers. The indecomposable data item is a bit. However, the basic data pathways and memory-addressing schemes for almost all general-purpose microcomputers address *groups* of *bits.* If this group has 4 bits, then the microcomputer is called a *4-bit micro.* A 4-bit group is usually called a *nybble.* If the basic bit grouping is in octets, or 8 bits to a group, then the microcomputer is dubbed with the adjective *8-bit.* Eight bits normally comprise one *byte.* The distinction between parallel and serial transfers derives from the method by which these bit groups are moved from the source to the destination. If they are moved 1 bit at a time over a single, 1-bit-wide path, then the transfer mode is said to be *bit serial* or *serial.* If the bit groups are moved a group at a time over multiple, parallel 1-bit-wide paths, then the transfer mode is said to be *bit parallel* or *byte serial.* Let us look at some examples.

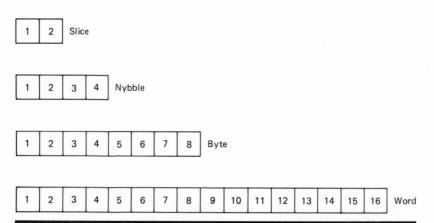

Fig. 2-2 *Bit groupings for microcomputers.*

Parallel Data Transfers

Microcomputer systems almost always transfer data that are to remain within the processor cabinet in parallel mode. In an 8-bit microcomputer system, all inside data transfers take place on an 8-bit-wide data path called the *data bus*. In addition to the eight data lines, at least two additional lines are required, a Signal Ground and a Data Ready line. A typical bit-parallel data path is illustrated in Figure 2-3. The *Signal Ground* line provides a reference point for determining the logic state of the eight parallel data lines, and the *Data Ready* tells the receiving device (destination) *when* to sample or read the eight data lines. The *when* function will be discussed in great detail in the section on controlling data transfers. Note that the eight parallel data lines are sampled simultaneously. Thus, you might say that the 8-bit data bus supports byte-by-byte transfers. Certain mainframe manufacturers have used the term *byte serial* to describe 8-bit parallel data transmission.

Another point to notice about the simple parallel data path shown in Figure 2-3 is that the control signals will support data transfers in *one direction only*. If the path is to support transfers in two directions, at least two additional control lines are required, namely an IN signal and an OUT signal to indicate which end of the channel is sending and which end is receiving.

Since the Z80 microprocessor is acting as the model 8-bit CPU for this book, let us look at the parallel data transfers that exist in a typical Z80-based microcomputer. First, within the Z80-CPU integrated circuit itself, there is an 8-bit data bus. This bus connects the internal CPU registers with each other, with internal CPU functions that interpret

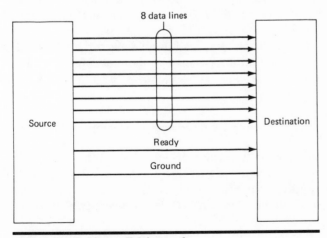

Fig. 2-3 Simple bit-parallel data path.

instructions, and with special-purpose logic to bring the content of the bus out to certain physical pins on the 40-pin package that houses the CPU chip. Z80 microcomputer memory is organized in 8-bit bytes. I/O ports transfer 8 bits at a time. The set of general-purpose registers supported by the Z80 CPU are 8 bits wide. It is worth mentioning that there are several Z80 registers that will hold 16 bits, and the memory-addressing scheme of the Z80 is based upon 16-bit addresses. However, each address and each 16-bit register is considered to be 2 distinct bytes that can be independently manipulated. In essence, the basic unit of data for the Z80 CPU is one 8-bit byte.

In the world of microcomputers, the width of the internal data bus is so important that it is usually the first characteristic used to classify a particular microprocessor. As we mentioned earlier, if the internal data bus is n bits wide, the microprocessor is called an *n-bit micro*. The impact of the data-bus width upon CPU performance is tremendous. A full explanation of this impact is outside the scope of this book; however, it is probably not hard to imagine the incremental speed and power of moving 16 bits instead of 8 bits for every memory access and for arithmetic and logic operations. The additional operation codes and associated single-instruction functionality are also significant. More than twice the benefit is realized by doubling the data-bus width. Popular 8-bit microprocessors that have been available for years are the Intel 8080/8085 family, the Motorola 6800 and 6809, the Mostek 6502, and the Zilog Z80 family. Sixteen-bit microprocessors have also been available for several years. The early members of the 16-bit group are the DEC LSI-11, the National Semiconductor IMP-16, and the Texas Instruments TMS9900. Several relatively new 16-bit processors have been developed: the Intel 8086, the Zilog Z8000, the Motorola 68000, and the National Semiconductor NS16032. Some of the latest 16-bit processors have a "few 32-bit characteristics." For example, the Motorola 68000 has a 32-bit-wide register set. This foreshadowed the announcement of the yet more powerful integrated circuits that implement full 32-bit CPU architectures and capabilities, such as the Intel iAPX 432 microprocessor.

Serial Data Transfers

External data transfers usually involve devices that are not housed in the same cabinet or are not in close proximity with the CPU and memory. Such transfers may occur over either parallel or serial interfaces. Since we discussed parallel transfers above, let us investigate the serial mode here.

A typical serial data path uses one line for the data-bit stream and possibly several others for Signal Ground and control purposes. There are several physical-level serial standards and conventions that specify a set of power and control signals. In Chapter 3 we will discuss three important serial-transmission conventions:

TTL level

EIA Standard RS-232-C

20-milliampere current loop

Each of these is widely used in microcomputer configurations for bit-serial data transfers between local components. RS-232-C is used for data transfers via the public-switched telephone network.

Parallel versus Serial Communication

Having discussed parallel and serial data transfers in microcomputer systems, let us look at how the two compare relative to several criteria.

Distance The distance of a parallel data transfer is usually less than 100 feet. Serial data transmissions quite commonly traverse distances from a few feet to thousands of miles.

Speed Data rates for parallel interfaces are potentially much higher over short distances, because multiple bits are transmitted simultaneously. Parallel data paths associated with typical microcomputer devices support data rates in the range from zero up to several million bits per second. Serial interface data rates associated with typical microcomputer devices range from zero to about 2 million bps. For both serial and parallel data transfers, the speed that can be adequately supported is inversely proportional to the distance of the transfer. Figure 2-4 illustrates this fact.

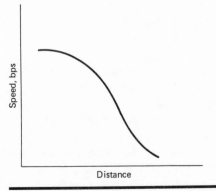

Fig. 2-4 How far versus how fast for data transfers.

Signal Levels Parallel interfaces usually employ transistor-transistor-logic (TTL) signals. That is, logic 1s and 0s are electrically represented on each parallel wire as plus or minus a nominal 5 V. Although TTL signaling is also used for serial interfaces, it is much more common to find serial devices operating at RS-232-C signal levels or at 20-mA current loop levels. All these types of interfaces are discussed in detail in Chapter 3.

Signal Loss and Amplification As an electrical signal is transmitted down a wire, there will always be some signal loss. This signal loss becomes more and more significant with the length of the wire. To compensate for signal loss, either higher-power transmitters or signal amplification can be used to preserve the signal's character. The amplification of one serial signal is significantly less complex than the amplification of multiple parallel signals. Phase and timing problems associated with the amplification of many parallel signals can become a significant cost factor. An important problem encountered with parallel transfers is skewing. *Skewing* occurs when the differences in individual line-propagation delays cause significant discrepancies in the timing of the individual data lines to be sampled by the receiver. As the distance increases, the skewing problems get worse.

Cost For distances over 50 feet, the cost of running multiple data lines becomes prohibitive. As we mentioned earlier, the design axiom "logic is cheap, communications is expensive" is particularly applicable to the relative merits of serial versus parallel transmission. The extra logic required to transform parallel data to serial data for transmission over a single data line, and then to reconstitute the parallel byte at the destination, is less expensive than the hardware required to effect a parallel transfer over long distances.

Telecommunications Up to the present time, the telephone network has served as the most common medium for long-distance communications. This network was designed for *analog voice* communications. Binary digital information that is to be transmitted over this medium must be serialized and, subsequently, converted to an audio signal (analog) appropriate to the telephone system's transmission capabilities. Thus, the public-switched telephone network constitutes a major facility for long-distance data communications for bit-serial data. Chapter 5 presents a detailed discussion of the characteristics of the telephone network with respect to data communications.

So, in comparing bit-parallel with bit-serial data transmission, one can conclude that each has applications for which it is best suited. For short-distance, high-speed transfers, parallel transmission is frequently preferable. For long-distance, lower-speed communication, bit-serial transmission is often the only alternative.

Whether the data are transferred over a serial or a parallel path, an important property of the transmission facility is its *direction*. The following paragraphs discuss the various types of data-transfer-path configurations in terms of their directional characteristics.

Direction of Data Transfer

Figure 2-5 illustrates three data paths with different directional characteristics. A simplex data path will support data transfers in only one direction, namely from device A to device B. Thus, A acts solely as a transmitter and B acts solely as a receiver in the first configuration.

The middle configuration illustrates a half-duplex data path, which will support *alternate* data transfers between device A and device B. In particular, device A can transmit to device B, and device B can transmit to device A, but these transmissions must not occur simultaneously. In this case, coordination must take place between the two devices at each

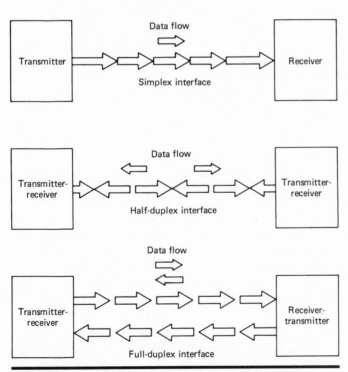

Fig. 2-5 *Simplex, half-duplex, and full-duplex serial interfaces.*

end of the half-duplex link to control line turnaround. This coordination takes the form of additional control signals that the interface must support.

The third configuration shows a full-duplex data path, which will support simultaneous data transmission in both directions. In particular, device A can transmit to device B at the same time that device B is transmitting to device A. Clearly, in order to implement this simultaneity, the resources for both transmission directions must be fully independent. Therefore, both devices A and B must have independent receivers and transmitters, and the data path from A to B must be totally separate from the data path from B to A. Thus, when A transmits to B and B transmits to A, essentially two logically (if not physically) independent simplex paths are being used.

Two terms that are frequently used in the context of data transfers over a communication bus are *unidirectional* and *bidirectional*. A *unidirectional bus* is a simplex data path that supports data flow from a particular source to one of several possible destinations. Thus, a *unidirectional bus* has only one transmitter and at least one, but possibly several, receivers. A *bidirectional bus* is a half-duplex data path that interconnects several potential transmitters with several receivers.

In microcomputer systems, the system bus is the vehicle that carries data between the CPU, memory, and peripherals. The lines that carry data constitute the data bus, which is bidirectional. The address bus is unidirectional, out from the CPU. Two additional classes of signals present on the system bus are bus-arbitration signals and system control signals. The bus-arbitration signals control who is master of the bus. In many microcomputer systems there is only one potential bus master, namely the CPU. However, it is equally common to have a second potential bus master that controls direct memory transfers, say between a mass-storage device (such as disk) and memory. The point we wish to make here is that the system bus is a very important resource in a microcomputer system that potentially can be controlled by one of several devices, examples of which are the CPU and a direct-memory-access controller. At any one point in time, the bus must have exactly one master. Thus, in systems where multiple potential masters contend for bus control, some control logic must exist to arbitrate bus requests and to grant the bus to the appropriate device. Bus request/grant logic and the associated bus signals are discussed in detail later in this section.

The bus master, once it has been granted the bus, uses the system control signals to coordinate: the direction of data flow, who can transmit, when transmission should start, when a receiver is ready, and other event-synchronization activities.

DATA-TRANSFER CONTROL

The previous sections discussed the two common modes of data transfer: parallel and serial. No matter what the mode, there must be some mechanism or *control* function that *synchronizes* the event sequence necessary for a successful exchange. Without synchronization, there are many ways a transfer is "bungled":

- The transmitter sends before the receiver is ready.
- The transmitter refrains from sending even though the receiver is ready.
- The receiver samples the data line(s) at the wrong time, thus reading garbage.
- Several transmitters try to send to the same receiver at the same time.

These are just a few of the possibilities. The solution to such problems is an agreed-upon procedure that governs the behavior of both the transmitter and receiver. Such an agreed-upon procedure involves the identification and definition of bus-arbitration and system control signals present on the system bus. The functions performed by these signals are so important that all data transfers may be classified by who controls them: the CPU or a direct-memory-access (DMA) controller. CPU-controlled data transfers are also called *program-controlled data transfers* or *transfers under program control*. Figure 2-6 illustrates these two control mechanisms for a data-transfer operation involving a CPU,

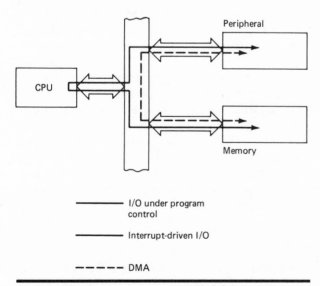

Fig. 2-6 DMA- versus program-controlled data transfers.

a peripheral device, and memory. Note that under CPU or program control, the path followed by data traveling between the peripheral device and memory passes through the CPU. In contrast, under DMA control, the data does not pass through the CPU.

Both the program-controlled and DMA-controlled data-transfer techniques use standard signals present on the system bus to specify the *where*, *what*, and *when* of the transfer.

Program-Controlled Data Transfers

Program-controlled data transfers transpire as a result of direct CPU execution of some I/O or memory-access instruction. Within the category of program-controlled I/O, there are two major techniques that are used by data-communications software: polled I/O and interrupt-driven I/O. We will describe the concepts underlying these two techniques here. Two experiments at the end of Chapter 4 provide examples of software and hardware necessary to implement polling and interrupt-driven device drivers.

Polling

In microcomputer systems, polling is sometimes employed to transfer data bytes between the console terminal and the CPU. In the case of bytes that are sent from the keyboard to the CPU, the traffic pattern is most irregular. Typically, the bytes come in bursts, with relatively long periods between some arrivals. In short, the CPU has no way of predicting when it will have to accept the next byte. The polling technique resolves this arrival irregularity problem by frequently examining the keyboard input port for an incoming byte. As long as the CPU queries the input port and accepts and processes incoming bytes faster than the keyboard produces them, no bytes will be missed.

Any worry that the CPU cannot keep up with the keyboard byte production rate is quickly dispelled when one considers the maximum typing speed of even the best typists. Suppose you could perfectly type 600 characters (bytes) per minute (this corresponds to 120 words per minute). In units of characters per second, 600 characters per minute converts to 10 characters per second. Thus, to keep up with the input character stream, the CPU would have to poll, input, and process one character every 100 ms. For a Z80 operating at 2 MHz, even a long instruction takes only 10 μs to perform. So 100 ms is enough time to execute literally thousands of instructions.

Therefore, quite the opposite is the case: The keyboard is not waiting for the availability of the CPU, but rather the CPU is waiting for the availability of the keyboard. As a matter of fact, the CPU is wasting perhaps 90 percent of its T cycles on "dry polls." Note, as the name implies, a *dry poll* is an unsuccessful poll, or a poll that finds no ready byte to transfer. Naturally, a successful poll is termed a *wet poll.*

An important performance issue associated with polling is the tradeoff of polling rate against the possibility of lost data. Clearly, by lowering the polling rate, one can reduce the waste due to dry polls. The CPU resources that are saved by a reduced polling rate can be utilized to perform productive work; however, this requires rather careful structuring of the tasks that the CPU is executing. For most microcomputer systems that use polling for data transfers, the CPU polls the I/O port at its maximum rate. Thus, for the case of keyboard input, the interval between character transmission is filled with as many dry polls as the CPU can generate in a tight polling loop.

If a CPU is using polling to exchange data with several peripheral devices, there must be some scheme for querying each device faster than its particular data rate to ensure that no data will be lost. The most common multiple-device polling scheme is the "round robin" technique in which each device is assigned a place in a (logical) circular arrangement and the polls go out to each device, one after another, around the circle. Prioritization of the devices in the round robin can be achieved by giving the highest-priority devices more than one polling slot in the circle. Figure 2-7 illustrates a round robin polling technique in which device 1 is given higher priority than the other devices in the circle.

Polling is normally used in systems where there are excess CPU resources that might just as well be used for polling, since otherwise the CPU would have nothing else to do. Therefore, single-user microcomputer systems quite often use polling for controlling data transfers between the CPU and the console terminal. As soon as timing or CPU resources become a problem, interrupt-driven I/O becomes the more desirable alternative.

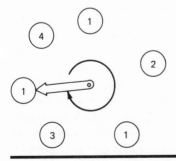

Fig. 2-7   Round robin polling technique.

Interrupt-Driven I/O

In interrupt-driven I/O, the peripheral device assumes a more active role in the data-transfer process by notifying the CPU when it is ready to accept or send a data byte. The following example serves to illustrate the difference between polling and interrupt-driven I/O.

Suppose you were expecting a telephone call from someone. If you were to employ the polling technique, your telephone would not be equipped with a bell and you would have to pick up the telephone frequently and ask, "Are you there?" If you were to employ an interrupt-driven data-transfer technique, you would equip your telephone with a bell and wait until you heard the telephone ring before picking up the receiver. Note that in the interrupt-driven case, the peripheral device (telephone) takes an active role in initiating the data transfer by generating an interrupt (the ring). Note also that the peripheral's interrupt capability requires some additional hardware to support it.

The Z80 microprocessor and most other microprocessors support two types of interrupt: maskable and nonmaskable. An interrupt is *maskable* if the CPU, under software control, can be instructed to ignore it. An interrupt is *nonmaskable* if the CPU may not under any circumstances ignore it. A peripheral device generates a maskable interrupt by activating a special CPU interrupt request signal called *INT**. To generate a nonmaskable interrupt, a peripheral must activate a special nonmaskable interrupt request signal called *NMI**.

How does a CPU service an interrupt? To answer this question, let us briefly look at the sequence of events that takes place during an interrupt-driven data transfer using maskable interrupts. First the peripheral device indicates its readiness by pulling INT* low. Assuming that the CPU has been programmed to acknowledge maskable interrupts, the CPU responds by executing a maskable interrupt request/acknowledge cycle. During this acknowledgment cycle, the peripheral device must identify itself so that the CPU will know what service it is to perform. In particular, for the Z80 CPU, the peripheral provides information that is sufficient to identify a special subroutine that contains the instructions associated with servicing the interrupting peripheral. This subroutine is called an *interrupt service routine*. Figure 2-8 illustrates the flow of control when an interrupt occurs.

The first task of an interrupt service routine is to save the state of the CPU so that processing can resume after interrupt service is complete. To ensure that the interrupt service routine will not be interrupted while saving the state of the CPU, the Z80 (and most other microprocessors) automatically disable maskable interrupts when an interrupt occurs. Once the state has been saved, interrupts may be made possible again with a software instruction. In the case of interrupt service rou-

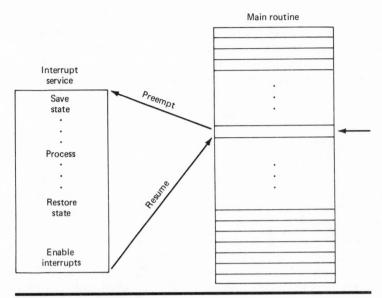

Fig. 2-8   Interrupt service.

tines for data transfers, the next task after saving the state of the CPU is to input or output the bytes to or from the peripheral. The last task is to restore the state of the CPU and "return from the interrupt." This last phrase essentially means that the CPU must terminate interrupt service and resume execution of the interrupted task from where it left off.

An experiment at the end of Chapter 4 provides an example of software for interrupt-driven I/O. In the experiment you will implement an interrupt service routine that will output data bytes to an external device (i.e., a line printer or terminal). As a background activity, the CPU will be continuously outputting characters to the console display.

Polling versus Interrupt-Driven Data Transfers

There are several positive attributes to the polling technique. First, as you saw from the telephone example given above, its associated hardware requirements are simple. Moreover, as you will see in the experiments at the end of Chapter 4, the software implementation of polling techniques is simple and the debugging of such software is straightforward. If the CPU has no other work to perform, polling is a way to utilize that resource for simplifying both the hardware and software associated with an application.

On the other side, the telephone example points out several disadvantages of polling. First, polling utilizes resources even though service is not required. Second, the wait-for-poll time associated with polling can be quite long. For example, in the telephone situation, if you were to sample your telephone at a rate of only once per hour, a caller might have to wait up to one hour before talking to you.

In contrast, interrupt processing allocates resources based strictly upon need. There is no such thing as a dry poll in interrupt-driven data communications. Therefore, the wasted resources associated with dry polls do not exist for interrupt-driven data transfers. In addition, interrupt service can begin immediately after the request is detected. Thus, the interrupt-driven data-transfer mode can achieve short delays prior to the commencement of service without incurring the overhead associated with frequent polling.

The actual time spent servicing an interrupt is usually longer than the peripheral service time in a polling environment. This is because an interrupt service routine must save and restore the state of the CPU in addition to executing the instructions directly associated with servicing the peripheral.

There are other disadvantages to interrupt-driven I/O. First, interrupt processing requires specialized hardware support. If multiple peripherals will be generating interrupts, circuitry must be provided to identify the interrupting device and to resolve potential conflicting resource demands. Note that device identification is no problem for polled data transfers because the CPU knows which device it is querying. Moreover, contention among multiple polled devices is automatically resolved by the order in which they are polled. In contrast, interrupt-driven data transfers require that both device identification and contention resolution be achieved in hardware. The most common contention-resolution technique for multiple interrupting peripherals is device prioritization. Typically, the devices are assigned relative priorities that define an "interrupt pecking order." For example, if devices 1, 2, and 3 are assigned priorities such that device 1 has the highest and device 3 has the lowest priority, then device 1 can interrupt service for both devices 2 and 3. Device 2 can interrupt service to device 3 but not to device 1. Device 3 cannot interrupt service to either devices 1 or 2.

Note that the above prioritization technique means that interrupt processing supports a preemptive service scheme. Since interrupts can arrive at any time, the CPU can be interrupted while servicing a prior interrupt. In the above example, if device 2 is currently receiving service and device 1 generates an interrupt, the CPU will suspend device 2's service and commence execution of device 1's interrupt service routine. Once the CPU completes service to device 1, it will pick up where it left off with device 2's service. In contrast, the most common polled I/O implementations support nonpreemptive service. That is, each device in the round robin is completely serviced before a poll goes out to the next device. However, polling implementations in which the polling software supports preemptive service do exist.

This leads to the second major disadvantage of interrupt-driven data transfers: logical complexity. The software required to support interrupt processing is significantly more complex than the software required to implement polling because interrupts are random events that can potentially interrupt a program at any stage of its execution. Therefore, quite often additional software logic is necessary to maintain the integrity of the state of the CPU and memory so that execution of an interrupted task can always be properly resumed. Another consequence of the randomness of interrupts is the occurrence of intermittent faults that are very difficult to reproduce, isolate, and repair. This tends to extend the time required for software development of interrupt-driven systems.

In the experiments at the end of Chapter 4 there are several examples of polled and interrupt-driven peripheral-device-driver software routines.

Direct-Memory-Access Data Transfers

Program-controlled data transfers, whether polled or interrupt-driven, utilize CPU resources to set up and perform the transfer. The direct-memory-access (DMA) technique implements the data-transfer function by adding a powerful new set of resources to perform the work, namely a DMA controller. Usually, the CPU is involved in the initial setup of the transfer, such as specifying the source and destination addresses, the number of bytes to transfer, and other related parameters, but the actual transfer operation is managed totally by the DMA controller.

To contrast DMA data transfers with the two types of program-controlled transfer techniques, let us extend the telephone example given in the last section. If you were to use a DMA technique to handle your expected telephone calls, you would hire an answering service to record your messages. Then you could call the answering service for the messages at your convenience. In a sense, the answering service constitutes a dedicated, specialized resource that you have added to augment your own limited resources. Moreover, when the answering service is using your telephone, that resource is not available to you. That is, the answering service controls the telephone until it decides to relinquish that control back to you.

In addition to achieving transfers independent of the CPU, an important characteristic of DMA transfers is that they are *block-oriented*. Typically, these transfers take place between memory and an I/O device. The overhead associated with setting up a DMA transfer is enough to make DMA inefficient for transferring just a few bytes. Almost always, DMA transfers move large chunks (at least several hundred bytes) of data.

Benefits of DMA

There are several benefits of DMA. One is that the CPU need not be tied up for long periods of time executing data-transfer software. In a DMA environment, the CPU sets up the DMA operation and then proceeds on to other work. When the DMA operation just set up by the CPU actually commences, the CPU must contend with the DMA controller for mastership of the system bus. Under some DMA techniques, once the DMA controller gets the bus, it keeps it until the entire transfer is complete. In this case, the CPU must enter an idling state because it cannot fetch instructions without control of the bus. Under other less-extreme DMA control techniques, the DMA and CPU interleave bus mastership, thus allowing the CPU to perform its work and the DMA to complete its transfer. In this last situation, both the CPU and DMA take a bit longer to complete their work, but at least the CPU is not tied up doing prolonged data transfers.

Some DMA controllers support a cycle-stealing, or transparent, mode of operation. In this mode, the DMA controller grabs the bus while the CPU is decoding the instruction that it last read. In this case, the only degradation in CPU speed is due to the continual changes in bus mastership, which does take some (but relatively speaking not much) time.

Another benefit of DMA is speed. As an example let us contrast the maximum data rates for Z80 program-controlled and DMA-controlled data transfers. A Z80 microprocessor operating at 4 MHz has a maximum program-controlled data-transfer rate of 0.2 Mbytes per second. This is actually very fast compared with other microprocessors. The reason that the Z80 CPU is so fast is that it has special block-transfer instructions which reduce the memory-read-cycle overhead associated with program-controlled data transfers. In contrast with the maximum program-controlled data-transfer rate of 0.2 Mbyte per second for the Z80, a Z80-DMA transfer (again with a 4-MHz system clock) can support a data rate of one Mbyte per second. With specialized external circuitry, this rate can be doubled.

A third, often less-recognized, benefit is reduced transfer response time. *Transfer response time* is the time that elapses between device readiness and the transfer of the first byte. This is especially true if initiation of a block transfer is in response to an interrupt from an I/O device. Even a relatively fast transfer of control to an interrupt service routine takes 5 to 10 μs. For many data-communications applications, such a long response will result in lost data.

Therefore, a computer system will be able to benefit from the performance advantages of DMA-controlled data transfers when any one of the following situations exists:

- The CPU is spending too much time executing data transfers to keep up with the rest of its workload.
- The data rates that must be achieved by a transfer operation are greater than the rates program-controlled I/O can support.
- The transfer response time must be faster than the transfer response time program-controlled I/O can provide without inappropriately elaborate start-up schemes.

Limitations of DMA

On the other side, DMA-controlled data transfers have limitations which can have significant negative impact upon system performance. These limitations derive from the fact that while a DMA transfer is in progress, the DMA controller is master of the system bus. In particular, when the DMA has the bus, the CPU cannot have it, which means that the CPU cannot fetch and execute instructions. If the CPU is responsible for managing dynamic memory refresh, a DMA operation will inhibit memory-refresh operations. Moreover, when the DMA has the bus, the CPU cannot detect and respond to interrupt requests from other devices in the system.

DMA transfers have two important sources of overhead. The first is bus access time, which is nonexistent for polled I/O. For interrupt-driven I/O with multiple devices contending for CPU service, the analog of bus access time is CPU access time. Since the DMA must compete with the CPU and possibly other potential bus masters for the system bus, there must be procedures for resolving contention for bus resources. These procedures take time, even with fast hardware to implement them. The second overhead source is associated with the setup of a DMA transfer. It is conceivable that a CPU would have to output two or three dozen bytes to the Z80-DMA controller to set up the DMA transfer. Thus, DMA setup can be significantly more time-consuming than initialization of a few registers for a program-controlled transfer. Therefore, if an application has short block lengths or requires frequent DMA reprogramming, the ratio of setup time to byte-transfer time may become unfavorable.

To summarize, when incorporating DMA capabilities into a system, one must stay aware of the consequences of DMA bus mastership on such time-critical functions as dynamic memory refresh and interrupt response. In addition, one should assess the overhead likely to result from bus contention and programming the DMA controller.

Applications of DMA

There are several data-transfer applications in which DMA can be especially suitable:

- *Hard Disk and Floppy Disk I/O:* The disk controller can use DMA to move blocks of data between the disk storage medium and memory.
- *Communications Channel I/O:* The interface between a computer system and a fast communications channel such as a local bus can use DMA to enhance response time, support higher data rates, and free the CPU to perform other work.
- *Multiprocessing and Multitasking Block Transfers:* For multiprocessor configurations, movement of data between private and shared memory is facilitated by DMA. For multitasking applications, paging and task swapping require the movement of large blocks of data, which benefits from the performance advantages of DMA.
- *Scanning Operations:* Any application that requires constant scanning of data blocks associated with some peripheral is a candidate for DMA. An example of such an operation is refreshing of a CRT screen. Screen- or block-oriented CRT I/O also can benefit from DMA.
- *Storage Backup Functions:* A common application for microcomputer systems is transfers between a Winchester fixed disk and streaming tape.
- *Data Acquisition:* When data arrives in large, dense bursts, DMA is often the only way to achieve the necessary response time and data-rate capacity.

These applications are only the ones most commonly encountered in microcomputer systems.

Hardware Support for DMA

Several semiconductor manufacturers market special-purpose integrated circuits that implement DMA functions. Since DMA controllers are designed and developed as dedicated, specialized components that are targeted specifically at data-transfer functions, they are typically configured to handle a broad variety of data-transfer functions. The DMA controller designed by Zilog for the Z80 CPU and other 8-bit microprocessors is one example circuit.

The data-transfer functions supported by the Z80-DMA controller may be summarized as three general classes of operation:

- Data transfers between a source (memory or I/O port) and a destination (memory or I/O port)
- Searches for a particular 8-bit target byte either in memory or at an I/O port
- Combined transfers with simultaneous searches

Each DMA product has unique features that differentiate it from its competitors. Other features commonly supported are multiple DMA channels per integrated circuit package, various techniques for sharing the system bus with the CPU, and timing options such as variable-width cycles.

The preceding discussion of data-transfer control techniques has concerned itself primarily with events on or near the microcomputer system bus. For data to be communicated from the CPU or DMA controller out from the computer cabinet to a local or remote peripheral, further control schemes and hardware are required. The communication link between the CPU cabinet and peripheral devices may transfer data in bit-serial or bit-parallel mode. In the last two sections of this chapter, we look at common techniques for implementing serial and parallel data transfers in microcomputer systems.

SERIAL I/O

Earlier in the chapter, we discussed the several advantages of serial data transmission. In many applications, its cost effectiveness more than justifies the complexities introduced by having to perform parallel to serial conversion on the parallel data that is indigenous to the inside of the computer cabinet.

There are two main problems that must be solved for two devices to successfully communicate over a serial link. The first problem relates to the mechanics of the serialization of parallel data to be transmitted and the conversion back to parallel form for received data. The second problem is mutual synchronization. The transmitter and receiver must coordinate their operation well enough to ensure that bits sent are sampled and read as bits received.

Parallel to Serial Conversion

Prior to the existence of medium- and large-scale integration (MSI and LSI), the parallel to serial conversion requirement made implementation of a serial I/O channel significantly more complex than implementation of a parallel channel. The function performed in parallel to serial conversion is primarily that of a shift register. For input operations, the shift register accepts single bits off the incoming line, moving the previous bits over one slot in the register, until all bits have been received. Figure 2-9 illustrates this process. The parallel to serial conversion function associated with transmitting data over a serial interface follows the reverse procedure, as shown in Figure 2-10.

Now there are specialized circuits that not only incorporate shift registers but also provide additional useful features such as direct connection to the system bus and programmable options for speed, parity sense, and character format. These fancy shift registers are marketed under several names:

- UART: universal asynchronous receiver-transmitter
- USART: universal synchronous/asynchronous receiver-transmitter
- SIO: serial input/output circuit

All these specialized parts use a combination of resynchronization detection logic (to recognize the beginning of transmission, a character, or a block) and shift registers either to accept a parallel byte off the data bus and shift it out, bit by bit, onto a serial channel, or to shift bits coming from a serial channel into an n-bit parallel-receive register that is interfaced to the data bus. The details concerning how to program the device and the options supported by the device vary according to the manufacturer of the component. Usually the technical manuals and

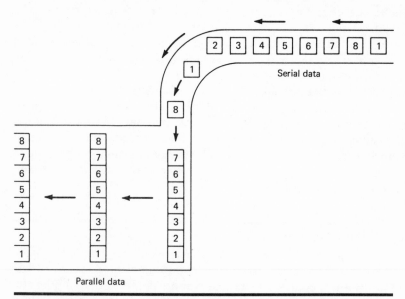

Fig. 2-9 *Receive serial to parallel conversion.*

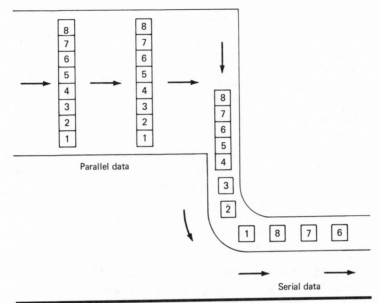

Fig. 2-10   Transmit parallel to serial conversion.

component data sheets for a particular serial I/O circuit provide good descriptions of this function, so we will not dwell on it here. The advent of UARTs, USARTs, and SIO circuits has greatly reduced the development time and, consequently, the cost of implementing serial I/O interface circuitry.

Device Synchronization

Two devices interconnected by a communications link are equipped only to discriminate between 0 and 1 logic levels on the line. Thus, if device A is applying logic 0 and 1 signals to the line (talking) and device B is actively sensing transitions on the line (listening), then it is possible to transfer data across the line. Notice that an important prerequisite to a meaningful data transfer is an agreed-upon encoding scheme that permits logic 0s and 1s to represent meaningful information.

In data-communications applications this encoding scheme usually represents characters such as letters, numbers, and punctuation marks as fixed-length bit strings. For microcomputer systems, the encoding method used most frequently is the ASCII standard. This standard is discussed in detail in Chapter 4. Hence, let us now assume that the source and destination devices have agreed upon a character encoding technique. There are still several issues that must be resolved between the correspondents before successful communications can occur:

- *Positive or Negative Logic:* Which of the two observable physical states represents a logic 0 and which represents a logic 1? For some communications links, two voltage levels on the line constitute the two observable physical states. For other interfaces, the physical states can be current or no current, two audio frequencies, or several other pairs. The issue for a TTL interface, for example, is settled by assigning either a logic 0 connotation to ground potential and a logic 1 connotation to nominal +5 V, or vice versa.
- *Bit-Transmission Order:* Which bit of a character is sent first?
- *Timing:* How do you know the difference between a series of identical bits and just an idle communications channel? How do you know when one bit ends and another begins? How do you know when one character ends and another begins?

All these issues are resolved by making an agreement and sticking to it. In the first two issues, implementation of an agreed-upon convention is straightforward. The first issue relates a physical signal to a logic value. As mentioned above, for computer applications, the physical signal can be a current, a voltage, or a frequency. Chapters 3 and 5 cover some of the common standards and conventions that map physical signals to logic values. The order of bit transmission is least significant bit first for the widely used serial-communication conventions. The most complex issue, timing, is discussed in the following section.

Timing

The first issue to resolve with respect to timing is to agree upon a bit-transmission rate or, equivalently, the duration of a single bit time. For example, two corresponding devices might agree that 1 s is equal to 1 bit time. Therefore, if device A asserts a logic 0 for 1 s, a logic 1 for 2 s, and a logic 0 for 3 s, device B would interpret what it sensed as the 6-bit sequence 111001, since the least significant bit is sent first. If there were some misunderstanding such that device B understood that 1 bit equaled ½ s, then it would have read device A's message as the 12-bit sequence 111111000011. Therefore, both the sender and receiver need to agree upon a bit-transmission rate, and, in addition, both must have a clock so that they can measure bit time.

A common problem with clocks is accuracy. Again, looking at a data transfer between device A and device B, suppose device A has a clock that registers the passage of 1 s every 1.05 s. Further, suppose that device B has a clock that registers the passage of 1 s every 0.95 s. From Figure 2-11, you can see that it may be possible to transmit 8 or 9 bits before this inaccuracy, or clock incompatibility, causes transmission

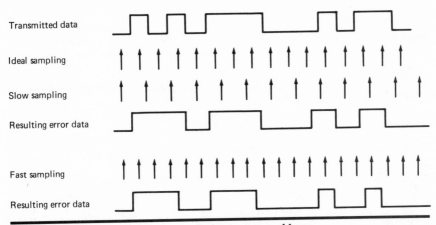

Fig. 2-11   Transmission errors due to clock accuracy problems.

errors. Clock incompatibilities can be compensated for by periodic resynchronization. Resynchronization is accomplished by several techniques, depending upon the specific transmission protocol. Clearly, the more incompatible the two clocks, the more often resynchronization must be performed. Since resynchronization is nonproductive in terms of sending messages over the communications channel, resynchronization essentially reduces channel transmission capacity. Bit times are used for housekeeping instead of data. So, an important trade-off in serial communications is between channel capacity and the tolerance of a channel to clock incompatibilities.

In addition to incorporating some tolerance for clock incompatibility the resynchronization mechanism resolves many other important issues identified above. In particular, resynchronization can be used to signify a transition from an idle to an active communications channel and boundaries between characters. Thus, the specific resynchronization or start-up mechanism is a major characteristic that distinguishes among serial link-level protocols. Two major classes of protocol are used in computer communications systems: asynchronous and synchronous protocols. The asynchronous is by far the most heavily used in microcomputer systems. Therefore, we will discuss the asynchronous protocol in great detail and will provide only a brief look at synchronous protocols. The next few paragraphs provide a quick overview of these two link-level protocols, and later sections discuss each in more detail.

Asynchronous versus Synchronous Transmission

In asynchronous serial transmissions, resynchronization occurs on a *character-by-character* basis. Resynchronization is coordinated by the use of start bits and stop bits that are collectively referred to as *framing* bits. The start bit represents a signal from the source to the destination that says in effect, "A data byte follows this bit." In synchronous transmissions, on the other hand, a separate clock signal is associated with the data. For local transmission, the clock signal may be carried over a separate physical line. For long-distance communications, the clock signal is encoded with the data at the transmission source, recovered from the data at the destination, and used to provide the necessary coordination between source and destination in order to accomplish a data transfer.

In addition to providing a separate clock signal, synchronous protocols provide for resynchronization on a data-block-by-data-block basis. A data block may be the equivalent of many data bytes or data characters. In fact, synchronous protocols are not necessarily character-oriented. That is, the management of data at this protocol level could be character-oriented *or* bit-oriented. Bit-oriented protocols transmit data as a bit stream with no character assignment imposed on the bits at this level of data-transfer management. It becomes the responsibility of a higher-level process to impose such an interpretation on the bit stream if necessary. Character-oriented synchronous protocols, on the other hand, recognize the existence of characters at this link level of data-transfer management. The clear trend now is toward the use of bit-oriented protocols.

In either case, asynchronous or synchronous transmissions, some framing bits must be added to the raw data bits in order to identify the beginning and end of characters in the asynchronous case and to identify the beginning and end of data blocks in the synchronous case. Thus, framing bits are added to each character for asynchronous serial transmission, whereas framing characters are added to blocks of data for synchronous serial transmission. These differences are illustrated in Figures 2-12 and 2-13.

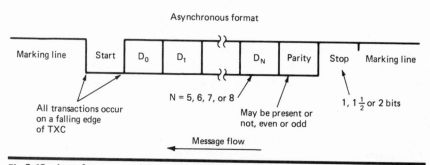

Fig. 2-12 Asynchronous serial character transmission.

Beginning   End

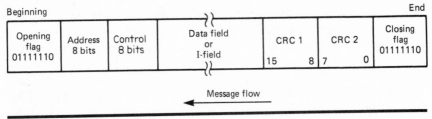

Opening flag 01111110	Address 8 bits	Control 8 bits	Data field or I-field	CRC 1		CRC 2		Closing flag 01111110
				15	8	7	0	

Message flow

Fig. 2-13   SDLC frame composition.

Asynchronous Serial Link-Level Protocol

In this section we will investigate the asynchronous serial link-level protocol in detail. As mentioned earlier, in asynchronous serial transmissions *characters* are transmitted at irregular intervals and the asynchronous serial data-link-level protocol resynchronizes on a character-by-character basis. Within a single character, the transmitter and receiver must measure 1 bit time to within at most 10 percent of each other. This 10 percent tolerance for transmitter and receiver clock differences arises from the fact that 10 bits quite often comprise one character, which means that after the tenth bit the faster clock will be exactly 1 bit time ahead of the slower clock. A difference between the two clocks of less than 10 percent implies that at least 10 bits can be sensed correctly before the clock inaccuracy causes errors. Thus, by performing resynchronization on a character-by-character basis, the asynchronous serial protocol incorporates some tolerance for inaccuracies between the transmitter's and receiver's clocks. Special character framing bits called *start* and *stop* bits are the mechanism through which character-by-character resynchronization occurs. The following paragraphs describe how a character is transmitted across an asynchronous serial communications channel.

Start Bit

Under the asynchronous serial protocol, an idle communications channel is at logic 1, or in a *marking* state. When a transmitter wishes to send a character, it first sends a start bit by placing a logic 0 on the line for 1 bit time. Thus, the receiver detects the transition of the channel from idle to active when the line goes low for 1 bit time. A potential problem is intermittent spikes that drive an idle line low for a very short time, thus creating the impression on the receiver of a start bit. Typically, receivers are equipped with spike-detection logic that samples the idle line at many times the data-transmission rate. In this manner, the receiver can sample an idle channel 2, 4, 16, or more times during 1 bit time. The spike-detection logic essentially helps to ensure the validity of a start bit by checking that the low line level does indeed last for an entire bit time. This mechanism is implemented by the software for the line monitor documented in Chapter 4.

In addition to synchronizing the beginning of a character, the start bit transition from a logic 1 to logic 0 state indicates the start of 1 bit time. Thus, the spike-detection circuitry, by sampling at several times the data rate, is able to recognize the start of a bit time to within a fraction of a bit time. For example, spike-detection circuitry that samples an idle line at 16 times the data rate can detect the start bit transition to within one-sixteenth of a bit time. Figure 2-12 illustrates the significance of the start bit as a synchronizing mechanism.

Data Bits

Once the start bit is validated by the receiver, the receiver sets up its shift register to begin accepting data bits off the line. While the number of data bits could be 5, 6, 7, or 8, usually either 7 or 8 data bits are transmitted. The number of possible characters (symbols) that can be represented with 5, 6, 7 or 8 data bits is calculated in the table below:

Number of data bits in a character	Number of possible characters
5	32
6	64
7	128
8	256

One can see that there is a possible problem with using 5 or 6 data bits per character. There are 26 uppercase letters in the alphabet, 26 lowercase letters, 10 digits, and at least 10 very commonly used special characters such as the following:

Special character	Symbol
Period	.
Comma	,
Semicolon	;
Colon	:
Dollar sign	$
Question mark	?
Parentheses	( )
Quotation marks	" "
Apostrophe	,

Thus it appears that a minimum of 7 data bits are required for a character set to support applications that involve text or word processing of any nature. However, the 5-bit *Baudot code* has been used since 1875, with *shifting* (figures shift and letters shift) characters employed to extend the basic 32-character set. Although the 5-bit Baudot code was implemented on the early teleprinters, it is gradually disappearing. The 7-bit ASCII code is the most universal character set in use on microcomputer systems today. ASCII stands for American Standard Code for Information Interchange. In fact, the ASCII code set is the most universal character set on systems of all sizes, not just microcomputer systems. Only the IBM 8-bit code EBCDIC (extended binary-coded-decimal interchange code) rivals ASCII in its widespread support. The ASCII character set will be discussed in detail in Chapter 4. The correspondence between binary bit patterns and the ASCII characters appears in a detailed discussion of ASCII in Chapter 4.

Parity Bit

Following the data bits a *parity* bit may be transmitted. The parity-bit convention is a mechanism to implement a limited error-detection scheme. The source knows that it has sent data to the destination, but there should be some mechanism to determine if the data was received by the destination in the same form that the source sent it. Albert Einstein is supposed to have said that life might not be very interesting without music. In some respects data communications might not be very interesting without errors. Noise and line interference produce data errors, and errors introduce all sorts of interesting and, at times, very complex problems into the world of data communications.

Various error-checking mechanisms can be implemented in order to reduce the probability that an undetected bit error exists between the data that the destination received and the data that the source intended the destination to receive. One method for detecting errors is for the source to provide the destination with redundant information. That is, the source provides the destination with the same information in two different forms. The destination can then compare the two pieces of information. If the two agree, the destination can assume that both pieces of information are correct. If the two pieces of information do not agree, the destination may conclude that something went wrong in the transmission process. The obvious method of providing redundant information is to send the same data twice. This is very wasteful. It either cuts the transmission rate of the channel in half or requires twice as many signal lines depending upon how it is implemented.

A more economical method of transmitting redundant information is usually employed. The data itself is *not* repeated, but a characteristic of the data is repeated. *Parity* is a characteristic of the data determined by the number of logic 1s that are contained in the data bits plus the parity bit itself. There are two choices. The parity bit can be determined such that the number of logic 1s on these bits is even or odd. Assume that the source and destination agree beforehand that the parity bit will be determined so that an *even* number of logic 1s exist on the parity plus data bits. In this case the scheme is called an *even parity* scheme. Alternatively and equally effectively, the source and destination could agree upon an *odd parity* scheme in which the parity bit is determined so that an odd number of logic 1s exist on these bits. Some examples of the calculation of the value of the parity bit are given below.

8-bit data byte	Parity bit even parity	Parity bit odd parity
0000 0000	0	1
0000 0010	1	0
1111 1111	0	1
1010 0110	0	1

A single parity bit can be used to detect single bit errors in the received data bits. In fact, parity will detect multiple data-bit errors as long as an *odd* number of them have occurred. Note that if an even number of invalid bits are received, the destination will not notice a parity inconsistency. For example:

Data byte transmitted	Data byte received	Even parity transmitted and received	Even parity calculated by receiver	Error detected
0000 0000	0000 0011	0	0	NO
0000 0000	1111 1111	0	0	NO
0000 0011	0000 0001	0	1	YES

Single bit parity is the error-detection scheme used with the asynchronous serial start-stop protocol. The calculation of the value of the parity bit to transmit at the source and the parity calculation and comparison at the destination can be accomplished in either hardware or software. Typically, asynchronous serial interfaces such as SIOs, UARTs, and USARTs perform these functions. If a discrepancy is detected, a flag, the parity error flag, is set. This flag can be read from a communications status register by the CPU and tested by the serial I/O device-driver software during the data exchange. The device-driver software may then take appropriate action upon detecting a parity error. The appropriate action may be a branch to a specialized error-handling subroutine, or, quite possibly, no action at all. For many applications, ignoring parity errors is entirely appropriate.

The frequency with which transmission errors occur, either detected or undetected, is called the *channel bit error rate* (BER). There are many factors that influence the channel BER: the type of cable, the type of line drivers and receivers employed, the environment as far as electrical noise is concerned, the installation techniques, the signaling method, and several other factors. However, for local transmissions that use off-the-shelf commercial components, the bit error rate is very small, and typically is smaller than 1 error for every 10^7 (10,000,000) bits transmitted. This is an average bit error rate that is equivalent to 1 error every 9 h or so of continuous transmission at a 300 bps data rate. Transmission errors usually occur in bursts. That is, many errors will occur in a very short time period, followed by no errors for a relatively long time period. This "burstiness" characteristic is especially common for transmission errors that occur across telephone lines. When the line takes a "hit," many errors occur in a short time frame. The typical bit error rate across telephone lines is approximately 1.5 errors for every 10^5 (100,000) bits transmitted. For a 300 bps transmission channel, this translates to approximately 16 errors for every hour of continuous transmission, or 1 error for every 4 min of continuous operation. It must be remembered that this is an *average* bit error rate. Since errors on the public telephone network tend to be quite bursty, the likelihood is great that an individual transmission will experience much different error rates (i.e., the standard deviation of experienced error rates is large).

The bursty nature of errors across a telephone line can be traced to the phenomena that cause the errors in the first place. The duration of interruption caused by lightning or by an electrically noisy switch in the switched telephone network is a relatively long time period compared with the bit time on even a low-speed communication line. A noise burst that lasts for one one-hundredth of a second (0.01 s) affects an average of 4 bit times on a 300 bps channel. Consequently, on a 300 bps channel, if one bit of a given character gets hit, the probability is quite high that at least one other bit in the same character will get hit.

There are many other error-detection schemes, such as longitudinal redundancy checks (LRCs), vertical redundancy checks (VRCs) and cyclic redundancy checks (CRCs), that are employed in data transfers. These schemes are frequently employed in synchronous communication links and are discussed in detail in some of the references. But for asynchronous serial data transmission, a single parity bit is the primary error-detection mechanism.

Stop Bits

Following transmission of the data bits and the parity bit, if one exists, the transmitter sends either 1, 1.5, or 2 stop bits. These stop bits are simply logic 1s that are 1, 1.5, and 2 bit times in length, respectively. Stop bits force the line to assume the marking condition for at least 1 bit time before the next character. This means that a start bit will begin with a transition from the logic 1 level to the logic 0 level. A stop interval was formerly required for mechanical coasting purposes in the old teleprinters. Typical asynchronous receivers manufactured recently do not require a stop interval for this purpose. However, at least one stop bit is necessary to ensure that each character begin with a transition from logic 1 to logic 0. This characteristic is important in the detection of framing errors, which are discussed later in this chapter.

Synchronous Serial Link-Level Protocols

In contrast with the character-by-character synchronization scheme used by the asynchronous serial link-level protocol, the synchronous communication scheme uses essentially two synchronization techniques. First, a sampling clock is used to define individual bits as they come across the line. This synchronization clock tells the receiver exactly when to sample the data line for incoming bits. There are two common methods for providing the data clock in a synchronous interface: "beside" or "within" the data. The "beside" data scheme uses a physically separate clock lead that carries the oscillating clock signal from the transmitter to the receiver. This scheme is quite acceptable for local communications, but it is somewhat impractical for long-distance transmission over the telephone network. Therefore, a "within" scheme exists that encodes the clock signal as part of the data signal. The device that performs this clock and/or data encoding is called a *synchronous modem*. Modems are discussed in detail in Chapter 5, so we will not address the clock encoding scheme here. Suffice it to say that a synchronous modem uses a signal-modulation technique that allows the receiver to recover the clock from the data it receives off the telephone channel. It then brings both the clock and data signals out on separate leads to the digital device attached to it. Thus, the digital device has no way of knowing whether it is communicating over a local or remote synchronous interface.

The second synchronization scheme used by the synchronous link-level protocols is designed to identify individual blocks of data. A block of data can be composed of any number of bits. As we mentioned earlier, the actual identification of groups of bits as characters may not occur until higher-level networking software decomposes and interprets the bits in the received data block. The synchronous protocols that do not identify bit groups with characters are called *bit-oriented protocols.* The protocols that do support specific character encoding schemes are called *character-oriented protocols.*

Synchronous protocols identify data blocks via special characters or bit sequences called "synch characters" in the case of character-oriented protocols, or "beginning flag" in the case of bit-oriented protocols. At the beginning of transmission or whenever a block resynchronization is to take place, a synchronous receiver enters the hunt mode, in which it searches the channel bit stream for the first occurrence of one or more synch characters. Upon receiving and identifying the requisite number of synch characters, the receiver may assume that it has "synch-locked" with the transmitter.

The following is a list of commonly used synchronous serial link-level protocols:

BISYNC (IBM)   Binary Synchronous Communications

SDLC (IBM)   Synchronous Data-Link Control

HDLC (ISO)   High-Level Data-Link Control

ADCCP (ANSI)   Advanced Digital-Communications-Control Protocol

X.25 (CCITT)   Recommendation from the X.25 Committee

DDCMP (DEC)   Digital Data-Communications Message Protocol

UDLC (UNIVAC)   Univac Data-Link Control

BDLC (Burroughs)   Burroughs Data-Link Control

Since serial microcomputer transmission is predominantly asynchronous serial, a detailed discussion of these protocols is not within the scope of this book. There are a number of excellent references that provide detailed information:

"Binary Synchronous Communications—General Information," IBM publication GA27-3004 File TP-09.

"IBM Synchronous Data Link Control—General Information," IBM publication GA27-3093 File GENL-09.

"ADCCP Standard: ANSI X3.66-1979," ANSI publication.

Overhead for Asynchronous and Synchronous Protocols

The asynchronous serial link-level protocol incorporates framing bits for each character transferred. The framing bits consist of 1 start bit, 1 parity bit, and either 1, 1.5, or 2 stop bits. Assuming that each character consists of 7 data bits, either 10, 10.5, or 11 bits must be sent to transfer such a character over an asynchronous interface. Therefore, if we assume that only one stop bit is used, only 7 of 10 transmitted bits are data bits. Fully 30 percent of the communications facility is consumed with framing information. This overhead rate remains constant, independent of the number of characters sent.

In constrast, the framing overhead for synchronous protocols arises on a block-by-block basis. The number of framing bits is the same for a long or short message. Therefore, the longer the message, the smaller the percent of communications resources consumed by protocol overhead. For example, let us compute the transmission overhead for SDLC. The composition of the basic SDLC frame is diagramed in Figure 2-13. There are a total of 48 overhead bits. For a message block consisting of only 8 bits, the SDLC protocol requires that 48 + 8, or 56, bits be sent for each 8 bits of data. Thus, only 8/56, or approximately 14 percent, of the communications link capacity would be used for carrying actual data. The remaining 86 percent would be dedicated to protocol overhead. The clear lesson here is that one should not use SDLC to send 8-bit messages. Let us look at a more reasonable message length, say 2048 bits. In this case, 2048/2096, or over 97 percent, of the transmission capacity is utilized for data bits—a considerable improvement. In general we have:

$$\text{SDLC Synchronous Protocol Overhead} = 1.0 - \frac{N}{N + 48}$$

where N represents the number of data bits transmitted. Overhead rates for other synchronous protocols can be computed using a similar technique.

Hardware Support for Serial I/O

As we mentioned earlier, there are several specialized integrated circuits that can be used to implement a serial I/O port. One of the earliest such parts to be commercially available is the UART (universal asynchronous receiver-transmitter). A UART not only performs the serialization and deserialization function but also inserts the framing bits associated with the asynchronous serial link-level protocol. A subsequent enhancement to the original UART is the USART (universal synchronous/asynchronous receiver-transmitter). As the name implies, a USART provides support for both synchronous and asynchronous communications. The functions performed by a USART operating in synchronous mode are detection of synch characters by the receiver and automatic insertion of synch characters by the transmitter when the CPU is not sending data. In this manner, the USART provides limited hardware support for data transfers over synchronous communication channels. The Z80-CPU family serial I/O interface circuit is the Z80-SIO circuit. Its capabilities represent significant enhancements over the earlier UARTs and USARTs in the areas of synchronous protocol support, event detection, and interrupt generation.

Error Detection

A very important function that is supported by virtually all serial I/O circuits is error detection. Three error conditions that can occur are:

Framing error

Parity error

Receiver overrun error

The sections that follow discuss each of these in terms of the Z80 SIO circuit's detection capability. The Z80 SIO circuit implements an internal status register that has individual bits designated as error flags for each of these error conditions. In a polling environment, the CPU may periodically read this register to check for errors. The SIO may be programmed to interrupt the CPU on the occurrence of these conditions to support interrupt-driven environments.

Framing Errors   Framing errors are detected when the receiver expects a stop bit and instead receives a logic 0 for that bit time. The stop bits are the only bits that the receiver can anticipate. The receiver knows that the start bit will be followed by a certain number of data bits, with or without a parity bit, followed by a given number of stop bits. If these stop bits do not all show up precisely when they are expected, the receiver knows that something is wrong. When a receiver detects an improperly framed data byte, a framing error condition results, and the receiver must assume that character synchronization has been dropped. Clearly, the most recently received bytes are suspect. Thus, a *framing error condition* may be defined as the absence of an expected stop bit. It could be the receiver's fault, the transmitter's fault, a hit on the transmission line, or whatever. The receiver can only report observation of an error; it cannot assign blame. Figure 2-14 illustrates a framing error condition.

Parity Errors   The receiver performs the parity calculation according to the previously agreed upon scheme (even, odd, or no parity) and compares the expected value of the parity bit to the actual value of the parity bit. If these do not agree, a parity error condition results. The SIO latches the parity error signal. Therefore, the SIO interface software must reset the parity error indicator once it has been triggered if a subsequent parity error is to be detected and recorded.

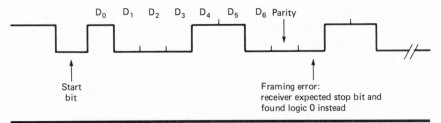

Fig. 2-14 Framing error condition.

The SIO may be programmed to generate an interrupt upon detecting a parity error. In this case, the interrupt service routine takes action that is appropriate to the particular application. This action could vary from doing nothing to collecting and computing detailed parity error statistics. Quite often the appropriate action is to initiate a retransmission procedure. Alternatively, if the destination (receiving) SIO port is operating in polled mode, the receiver-driver software frequently does not check for a parity error on a character-by-character basis but rather on a line-by-line or block-by-block basis. In the block-by-block or line-by-line case, detection of a parity error implies that the entire line or block has an error somewhere. From the single-bit parity error indicator, it is impossible to identify the errant character(s). Therefore, the entire line or block must be retransmitted.

In a polled environment, there is clearly a trade-off between incurring the overhead of checking for a parity error on a character-by-character basis and recovering errors on a block-by-block basis. On a character-by-character basis, once an error is detected, the recovery of only one character is necessary. However, time and resources are spent examining each character to determine if there was a parity error. On the other hand, there is comparatively little overhead incurred in checking for a parity error on a block-by-block basis, but once an error is detected, the entire block is questionable and may have to be recovered.

Receiver Overrun Errors   Asynchronous serial receive interfaces are typically capable of buffering a few characters. The Z80-SIO, for example, buffers three characters on the receive side. The buffer is organized as a FIFO (first in, first out) queue. The buffer "overflows" when four successive characters have been received by the SIO without being read and consequently removed from the buffer. When this overflow condition occurs, the retention philosophy for the SIO is to retain the last character received at the expense of the second-to-last character received. The notion of a FIFO buffer and this retention philosophy are depicted in Figure 2-15.

External Status Conditions

An additional capability of the SIO circuit is to monitor and report the status of certain external conditions, such as modem status and BREAK condition status.

Modem Status If the SIO is interfaced to the public-switched telephone network through a modem, the status of the RS-232-C signals Data Carrier Detect and Clear to Send is critical because they describe the modem's and channel's readiness to transfer data. Both these signals are discussed in detail in Chapters 3 and 5. So, for now, we will merely acknowledge their importance and state that the SIO latches their state in a status register. The CPU may poll this register or program the SIO to interrupt it on any change in status of either signal.

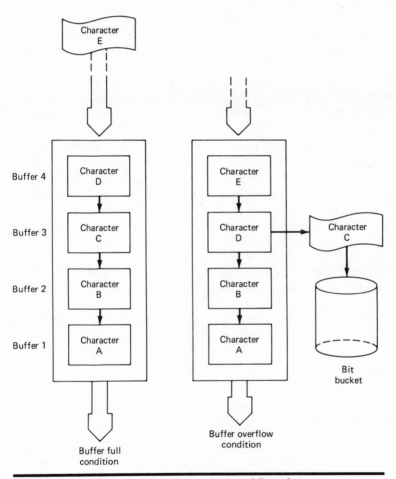

Fig. 2-15   SIO/0 receive FIFO buffer & retention philosophy.

Break Condition The second external status condition monitored by the SIO is the existence of a BREAK condition. The following paragraphs discuss what a BREAK signal is and then summarize SIO support for handling of BREAK conditions.

Virtually all serial terminals have a key called a break key which is usually labeled BRK or BREAK. This is a very special key. The key does not represent an ASCII character but rather a special signal to be transmitted from a terminal to a host computer system. The purpose of the break key is to provide a method for a terminal to send a special signal to the host system at any time. The action that the host system takes upon the detection of such a signal depends on the particular system involved. However, the BREAK signal is usually used to indicate that the terminal user wishes to take some relatively drastic action, such as suspend the current mode of operation and initiate a new mode of operation, or abort the process currently in execution.

The break key generates a *signal*, not an ASCII character. The BREAK signal, by definition, occurs when a transmit line is held at logic 0 for a certain minimum time period. Depressing the break key has the effect of holding the transmit line at logic 0 for the duration of the time period that the key is depressed. Once the key is released, the transmit line returns to the logic 1, or marking, state. No stop bits are generated during the time period that the break key is depressed, so the destination computer system simply perceives a terminal line that is being held at logic 0. A *minimum* time period is established for each individual computer system that is used as the definition for the length of a valid BREAK signal. While this minimum time period is a system-dependent characteristic, typical minimum time values fall between 100 and 600 ms. Thus, if the Transmit Data line is held at logic 0 for a time period that exceeds the minimum threshold, the system will detect a BREAK condition and respond accordingly.

The Z80 SIO has provisions for both generating and detecting a BREAK signal. Many other serial interfacing components have similar capabilities.

There are four important functions associated with receiver recognition of a valid BREAK condition from a terminal:

- Break signal detection
- Break signal termination
- Break signal timeout
- Break signal validation

The first two, detection and termination, have SIO support in the form of associated status conditions or interrupts. The BREAK signal timeout function requires a real-time clock. The last function, validation, relates to the sequencing in time of the other three. For additional information on the exact mechanisms used to perform these functions, we refer you to

Zilog *Z80-SIO Technical Manual and Application Notes.*

Nichols, Nichols, and Musson, *Z80 Microcomputer Advanced Interfacing*, Book 3, Howard W. Sams, 1982.

PARALLEL I/O

The last topic that we discuss in this chapter is parallel I/O. We have saved the simplest I/O technique for last. The reason that parallel I/O is so much simpler than serial I/O is due to the removal of several complications that arise from the parallel to serial conversion requirement. The next section identifies several functions that are desirable for support of parallel interfaces.

Functional Characteristics

Most microcomputer parallel I/O ports implement eight data lines plus one or more "handshaking" lines. Figure 2-16 shows a very simple 8-bit parallel I/O port with one handshaking line, labeled READY. The term *handshaking* in data communications refers to the synchronization and coordination that must occur between communicating devices. For the particular parallel port in Figure 2-16, the Ready line makes a transition from logic 0 to logic 1 to notify the receiving device that all 8 data bits are present on the 8 data lines. The receiving device monitors the Ready line, and, when the transition occurs, it reads the logic levels present on the parallel lines as the byte that the source intended it to receive. Several points are worth noting about this simple parallel interface.

Character Synchronization The Ready signal synchronizes the reception of characters by the destination. The transmitter totally controls the parallel interface. If the receiver cannot keep up, there is no way for it to notify the transmitter. Therefore, most parallel interfaces implement one more handshaking signal, a Strobe line. The Strobe line is controlled by the receiver to notify the transmitter that it is ready to

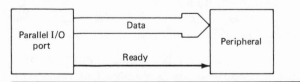

Fig. 2-16 Simple parallel interface.

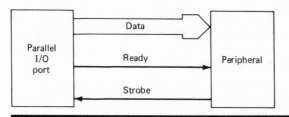

Fig. 2-17   Parallel I/O interfacing with two handshaking lines.

accept the next character. Thus, the transmitter monitors the Strobe line, and only sends bytes when the receiver activates the Strobe line. Figure 2-17 shows a parallel I/O interface with both a Strobe and Ready line for handshaking.

Signal Names   Unfortunately, in the computer-equipment-manufacturing industry, the names for the above parallel I/O handshaking signals are not standardized. Therefore, in some contexts you will see different, and often confusing, naming conventions. The most confusing convention that we have seen occurs in the context of parallel printer interfaces. These interfaces implement two handshaking signals called *Strobe* and *Busy/ Ready.* The function of the Strobe signal is analogous to the function of the Ready signal that appears in Figure 2-17. The function of the Busy/Ready signal is analogous to that of the Strobe signal shown in Figure 2-17. You can see now why it is easy to get confused. The only way to be absolutely sure that you understand a parallel interface (to a printer or some other peripheral device) is to read beyond the signal names, look at the schematics provided with the equipment (to see signal direction, if nothing else), and carefully read the descriptions of the signal functions. In this book, we are using the signal name and signal function convention that is illustrated in Figure 2-17. This convention seems to be fairly universal among semiconductor manufacturers. The other convention mentioned above is more commonly found in the vocabulary of computer peripheral manufacturers. Throughout the book, our signal-naming conventions remain consistent with our model-Z80-based microcomputer system.

Overhead   The concept of overhead, as defined for serial communications, does not exist for parallel interfaces. Each bit carried by the eight data lines is indeed a bona fide data bit. However, there is at least one, and most likely two handshaking lines. Thus, overhead, in the context of a parallel interface, is incurred in the form of extra signal lines.

Error Detection   Error detection in a serial communications channel is achieved through sending redundant information. This redundant information is contained, for the asynchronous protocol, at the end of each byte, in the form of a single parity bit. For synchronous protocols, more bits are dedicated to the error-detection function through the use of multiple bytes (e.g., the cyclic redundancy check bytes of the binary synchronous communications protocol and others) appended at the end of a message block. Thus, for serial I/O, error checking increases transmission overhead. The inalterable fact is that there is no free lunch. Hence, a parallel interface pays for error checking in the currency of its overhead costs, namely extra transmission lines. For a parity bit, a parallel interface typically implements one additional lead, called a *parity signal*. The transmitter, based upon an agreed-upon even or odd parity scheme, computes what logic level to place on the parity line. The receiver performs the same computation based upon the data it received, makes the comparison, and duly hiccups if something is awry.

Hardware Support for Parallel I/O

The Z80-PIO circuit, like the Z80-SIO component, is a programmable, multipurpose communications interface chip. The functions and special capabilities of the Z80-PIO circuit are documented in the Zilog Z80-PIO Technical Manual. Additional detailed PIO experiments and application notes can be found in:

Nichols, Nichols, and Rony, *Z80 Microprocessor Programming and Interfacing, Book 2,* Howard W. Sams, 1978.

Nichols, Nichols, and Musson, *Z80 Microcomputer Advanced Interfacing, Book 3,* Howard W. Sams, 1982.

CONCLUSION

This concludes what has been a rather lengthy general discussion of data transfers. We have discussed terminology, transfer control schemes, synchronous and asynchronous communications protocols, and hardware support for serial I/O, parallel I/O, and direct memory access. The underlying example for all of this discussion has been a Z80-based microcomputer system. We wish to emphasize that, while many of the detailed mechanisms presented here are specific to Z80-based systems, the principles certainly are not. The remainder of the book continues in this vein. Specific examples and designs are introduced that, on the one hand, lend concreteness to the ideas presented and, on the other hand, introduce some Z80 dependency or processor-unique aspects to the book. We continually try to strike a reasonable balance between the two, so that both the general principles and real-world realization of concepts are conveyed.

3

RS-232-C and Other Physical-Layer Protocols

In this chapter, we wish to address the problem of physically transferring data between electronic devices. Associated with a data transfer are a source, a destination, and a connection between them. Consider the problem of communication between two integrated circuits on the same printed circuit card, for example, a memory-read cycle executed between a CPU and some on-board memory. In this case, the data travel over metallic printed circuit pathways that have been etched into a plastic board.

Typically the data representation is binary, where a nominal +5 V is equated with logic 1 and a nominal 0 V is equated with logic 0. This is called *TTL* (for *transistor-transistor logic* level) signaling, and it is the standard technology for component-to-component communications inside a computer cabinet. It is impractical to define a standard that requires *exact* voltages, so, in actuality, TTL associates voltage *ranges* with the two logic conditions. These ranges are shown in Fig. 3-1. Note that the voltage ranges for devices that are *sending* are slightly different from those for devices that are *receiving*. Thus, a *sender* (sometimes called transmitter, driver, or generator) must assert a voltage of at least 2.4 V to transmit a logic 1. However, the *receiver* (sometimes called terminator) requires only a voltage in excess of 2.0 V to sense a logic 1.

The difference allows for some loss in voltage between the sender and receiver. The allowance is called a *noise margin*, which, in this case, is 0.4 V for sending either a logic 0 or logic 1. As you will see, the practice of differentiating between sender and receiver voltage-to-logic-level correspondence is quite common in specifications for the electrical characteristics of an interface standard.

Also note that Fig. 3-1 shows the existence of a relatively large voltage range for which no logic condition is defined. This region is somewhat smaller for the receiver. Thus, the receiver's voltage-level standards for representing a given logic condition are somewhat relaxed in comparison with those for a sender.

For internal computer data transfers, TTL-level signals are ideal for several reasons. First, their associated power requirements and heat dissipation are low. Second, TTL-level signals interface directly to integrated circuits without the need for costly line driver and receiver circuits. Finally, TTL interfaces operate at the high speeds that are required for internal computer data transfers.

What about communication between distinct digital devices, such as a computer and a local terminal or two computers in the same building? It was natural to try to extend the techniques that worked so well inside the computer cabinet to data transfers between cabinets. This has been done, and today there are several commercially available terminals that support TTL-level signaling with a host computer. Unfortunately, there

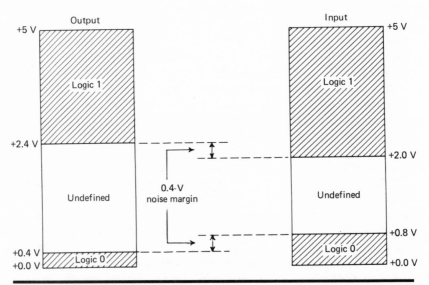

Fig. 3-1   TTL interface electrical characteristics.

are several very serious problems associated with TTL communications over distances of more than a few feet. First, TTL is very susceptible to externally induced noise. Second, line losses that reduce the voltage potential of transmitted signals have a particularly significant impact upon the 0- and 5-V TTL levels in that losses of even just a few volts can totally obscure the difference between a logic 0 and logic 1.

TTL-type communications are most often carried out via parallel data transfers. Parallel data transmission is inappropriate for distances over tens of feet for at least two important reasons: reliability and cost. Reliability is adversely affected by the phasing and skewing problems that are likely to develop in parallel interfaces; and the cost of cable and connectors to support parallel data transfers is higher than that for serial-mode communications.

The logical extensions to data transfers between two devices in the same room (separated by tens of feet) are data transfers between devices in different buildings, different cities, different regions, different continents, and finally . . . different planets. Each of these types of communications presents special problems that must be solved using techniques and technologies that are appropriate.

In the late 1940s, when the first electronic computers were being developed, the Bell System was the primary nonmilitary source of long distance communications expertise. It did not take long for computer systems designers to realize the potential of the existing Bell network for communication of data between remote computer sites. In 1950, the U.S. Department of Defense, IBM, and AT&T embarked upon the first major project that involved sending digital data over the telephone network. This project, called SAGE, for Semi-Automatic Ground Environment, was designed for North American air defense. Radar data accumulated at geographically distributed remote sites were sent at 1200 bps, over dedicated telephone links to centralized computer sites.

From 1958, when the first SAGE sector became operational, until 1967, all usage of the telephone network for data communications was via devices called modems that were hard-wired to the telephone lines. A user wanting to relocate a computer terminal from one room to another room had to pay a telephone company modem installer to come out to the site and physically move the data line and modem to the new location.

Early in 1967, John Van Geen, an engineer at the Stanford Research Institute, built the first portable modem. Since the Stanford Research Institute was not in the manufacturing business, it made Van Geen's design available to private industry. Reid Anderson, who had just started a new company, called Anderson-Jacobson Corporation, saw the design and decided to commercialize it as a product for the industrial market. In August of 1967, the first commercial acoustic (portable) modem was sold.

The significance of this device was that it provided a convenient, low-cost connection to the vast resources of the worldwide telephone communications network. The telephone company had been working on the problem of long-distance voice communications for years, and, fortunately, their existing plant would support communications between digital devices. The problem of converting ON/OFF binary signals to a form that could be effectively propagated through the telephone network was solved inexpensively and conveniently by using the portable acoustic modem. A new problem was created, however. A standard was now required that would address the interface between the communicating digital devices and the modem. In 1969, the Electronics Industry Association (EIA) published the standard that is now used in virtually all microcomputer systems and larger systems as well. This standard, usually known as RS-232-C, is the major topic of this chapter.

OBJECTIVES

When you have completed this chapter, you will be able to do the following:

- Identify and define the term *layer communications protocol.*
- Describe several standards and/or conventions for physical-layer protocols that commonly appear in microcomputer environments: TTL, 20-mA current loop, and RS-232-C.
- Understand the function and significance of each signal implemented in a standard RS-232-C full-duplex cable.
- Describe the RS-449 standard that the EIA and other standards organizations propose to replace RS-232-C.
- List the relative strengths and weaknesses of TTL, 20-mA current loop, RS-232-C, and RS-449.
- Design and build RS-232-C cables that will support serial communications for your microcomputer system.
- Build and use simple tools that can be useful in cable design and communications troubleshooting over RS-232-C serial links.

PHYSICAL-LAYER COMMUNICATIONS PROTOCOLS

A physical-layer communications protocol sets forth rules concerning mechanical connections, electrical signal characteristics, and signal functional characteristics that must exist for the interface to be in compliance with the standard. The ultimate objective is for two distinct equipment manufacturers to be able to build their respective devices by following the standard and come out in the end with full compatibility, i.e., two devices that can effectively exchange data. Sets of rules that address the purely physical aspects of transmitting electrical signals between communicating devices are called physical-layer communications protocols, the subject of the next section. Other, higher-level aspects of protocols for data communications are discussed in subsequent chapters.

Data are transferred between electronic devices via some type of interface that consists of electrical impulses, cable to carry the impulses, and connectors that attach the cable to the devices. The data are commonly represented by changes in current or voltage. To accomplish successful data interchange, devices must follow a physical-layer communications protocol. Such a protocol sets forth standards that unambiguously resolve the following issues:

- *Mechanical Characteristics:* Detailed specifications are given for the connector dimensions, number of pins, pin assignments, diameter of pins and sockets, connector location, and cable characteristics such as length and number of conductors.
- *Electrical Signal Characteristics:* The electrical characteristics of the data interchange signals and the associated circuitry must be specified. This includes a specification of maximum data rates, identification of voltage or current levels that represent signal status conditions (logic level, ON/OFF, MARK/SPACE), and specification of characteristics of the receiver and transmitter circuits.
- *Functional Description of Signals:* The signals that comprise the interface are usually characterized by function, by whether they are from originator or receiver, and by their relationship with other signals.

In this chapter, we will limit our discussion to the physical-layer protocols that are most common in microcomputer-based systems. First, we will take an in-depth look at some serial-transmission protocols. These include several standards that are defined by EIA as follows:

RS-232-C "Interface between Data Terminal Equipment and Data Communication Equipment Employing Serial Binary Data Interchange"

RS-422 "Electrical Characteristics of Balanced Voltage Digital Interface Circuits"

RS-423 "Electrical Characteristics of Unbalanced Voltage Digital Interface Circuits"

RS-449 "General Purpose 37-Position and 9-Position Interface for Data Terminal Equipment and Data Circuit-Terminating Equipment Employing Serial Binary Data Interchange"

Our investigation of serial communications will concentrate primarily upon the RS-232-C standard. However, since the other three standards are intended to gradually replace RS-232-C, they will undoubtedly become quite important in the near future. Since RS-422, RS-423, and RS-449 are compatible extensions of RS-232-C, our emphasis in considering them here will be to make a comparison and to provide information that will facilitate the transition to these new standards.

The last serial interfacing convention that we discuss is the 20-mA current loop. Although no universally accepted standard is defined for current loops, they are still in widespread use.

The second major category of interface conventions relates to parallel communications. These interfaces commonly occur in interfacing printers to microcomputers. Unlike the serial standards mentioned above, the parallel conventions are defined by specific vendors of printers and specialized parallel I/O port circuits. The parallel handshaking conventions that were discussed in Chapter 2 are examples of common parallel interface practices.

SERIAL COMMUNICATIONS: RS-232-C

Background

Anyone who has read an advertisement for a computer terminal or printer has probably seen the phrase "RS-232-C compatible." These words almost always mean that the device can be connected to a computer with a specific 25-pin connector and that the electrical and mechanical characteristics of the interface *do not violate* the RS-232-C standard promulgated by EIA. What should *not* be assumed from these words is that *all* the RS-232-C signals and functions are supported, for the vendor has most likely implemented only a subset. It is important to understand the various signals present in the RS-232-C standard so that you can read vendor specifications and judge if the RS-232-C subset that has been implemented is indeed appropriate for your requirements. The following is a summary of the RS-232-C standard. Selected portions of the published standards document have been reprinted with the permission of the Electronic Industries Association, for which we give grateful acknowledgment.

Need is the mother of invention, and the need from which RS-232-C was born arose from the increasing utilization of the telephone network for data communications. In the United States, the Bell System operating telephone companies were the first major suppliers of data communications service, and as such, were the first major suppliers of equipment capable of interfacing digital devices to the telephone network. This equipment, the Bell modems, which were developed at Bell Laboratories, became the industry standard, and to a significant degree, they still are today. We will discuss modems in detail in Chapter 5. As more and more computer equipment manufacturers began to develop terminals, computers, and other devices that could interface to Bell modems, the need for information about the modem interface requirements grew. In response to this need, the EIA, the Bell System, and independent modem manufacturers, in a joint cooperative effort, developed the RS-232-C standard.

The RS-232-C standard was published by EIA in 1969. The letters RS stand for Recommended Standard. The 232 is an identification number, and the final C indicates how many revisions the standard has gone through.

As the full title of the RS-232-C publication indicates, the standard addresses communications between data terminal equipment (DTE) and data communication equipment (DCE). The initials DCE now stand for data circuit–terminating equipment. The significance of this fact to your achieving successful RS-232-C data exchanges with your microcomputer system is questionable. However, it will assist you in understanding much of the literature that is currently published. The formal definitions of the terms DCE and DTE below are taken from the glossary of *Technical Aspects of Data Communication*, by John McNamara (Digital Press, 1977).

DCE: The equipment that provides the functions required to establish, maintain, and terminate a connection, and to perform signal conversion and coding required for communication between data terminal equipment and data circuit. The DCE may or may not be an integral part of a computer.

DTE: (1) The equipment comprising the data source, the data sink, or both. (2) Equipment usually comprising the following functional units: control logic, buffer store, and one or more input or output devices or computers. It may also contain error control, synchronization, and station identification capability.

Figure 3-2, which shows DCEs and DTEs in a communications circuit, illustrates the major points of the above definitions.

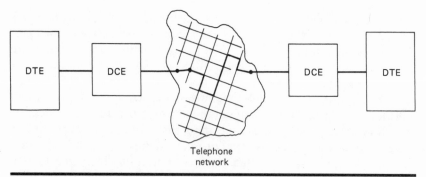

Fig. 3-2   DCEs and DTEs in a communications circuit.

Essentially, the DTE represents the ultimate source and/or destination of the data. Examples of DTEs are receive-only printers (data destination) and CRT/keyboard data terminals (either data source or destination). The DCE facilitates the communication of the data from its source to its destination. A modem is a device that is a DCE. Let us look at some examples.

Example 1   Consider the configuration illustrated in Fig. 3-3 in which a CRT terminal is connected to a remote time-sharing host computer over the telephone lines. Two DCEs and two DTEs are shown. The DTEs are the CRT terminal and the remote host because they are both ultimate senders and receivers of data being communicated. The two modems at each end of the telephone connection are DCEs because they perform the signal conversion necessary for achieving communications over the switched public telephone network. The data-communications circuit that the two DCEs terminate is the telephone link.

Note that the RS-232-C standard governs communication links

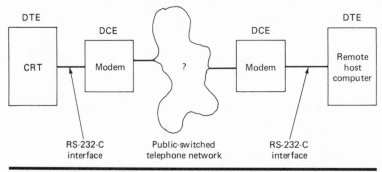

Fig. 3-3   Example 1 configuration.

between DCE and DTE and has nothing to say about the telephone net-work connection between the two DCEs. Thus, the telephone communications network is treated like a giant "black box." Essentially, you, as a designer of systems with remote communications requirements, need not concern yourself with the particular route that your data traverses through the telephone network. However, there is one important caveat. You certainly will be concerned about the *performance characteristics* of that black box and they will definitely be a function of the route that your data travels. As an example of the dependency of performance upon the routing of data through the telephone network, consider the difference in propagation delay for terrestrially routed messages versus messages routed by satellite. On the average, the one-way travel time via a terrestrially based route can be approximated at 1 ms per 100 miles. Because of alternate routing of long-distance switched connections, this can vary by as much as 20 ms, but it is a reasonable average. On the other hand, a typical communications satellite is in geosynchronous orbit at about 22,000 miles above the earth's surface. Each "big hop" (satellite link) in a message's route adds 700 ms to the travel time. Where this can become critical is in designing interactive systems in which the remote host echoes each character as it is typed by the user at a keyboard. If there is a satellite link in the path from user to computer, the round-trip delay is a minimum of 1.4 s (1400 ms) just to echo a character. Clearly this is totally unacceptable.

Example 2 Consider the configuration shown in Fig. 3-4, in which a CRT terminal is connected to a local microcomputer via an RS-232-C interface. Which are the DCEs and the DTEs? The old aphorism that it takes two to do the RS-232-C tango applies here. RS-232-C supports communications between a DCE and a DTE. Since there are two "equipments" in the subject configuration, they must at least *appear* to each other to be a matched pair. Typically, CRTs and other data send/receive terminals are configured as DTEs. Some microcomputers are configured as DCEs and some as DTEs. A Cromemco's I/O ports are all configured to be DCEs. The serial I/O ports of a TRS-80 look like DTEs.

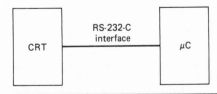

Fig. 3-4 Example 2 configuration.

What about your computer? You will determine this later in an experiment.

What happens if both your terminal and computer are DCEs? Similarly, what happens if both your terminal and computer are DTEs? In both cases you have a problem. Luckily it is one that can be easily solved by a simple cable called a *null modem*. We will discuss null-modem cables in detail later, but let us briefly look at their function now. The name is quite suggestive of the function. First, the word *null* implies that it doesn't do anything, at least nothing significant as far as the form and content of the data are concerned. Second, the word *modem* implies that it is a DCE. Thus, for two DTEs, the null modem constitutes a DCE that sits between them and resolves the requirement for a DTE-DCE RS-232-C interface pair.

What all this boils down to is the following two crucial facts:

1. There are two lines that carry data between a DTE and DCE, one for each direction.
2. The DTE sends on line A and the DCE receives on line A. Similarly,the DCE sends on line B and the DTE receives on line B.

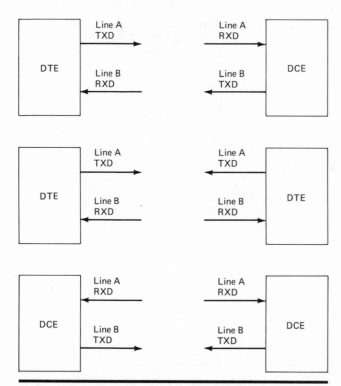

Fig. 3-5   DCEs and DTEs transmit and receive lines.

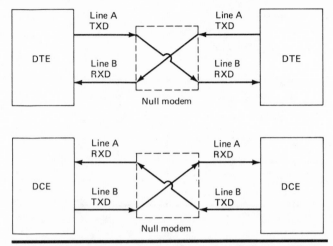

Fig. 3-6 *The functions of a null modem in interfacing two DTEs or two DCEs.*

Figure 3-5 illustrates the above points. Consider what happens when two DTEs attempt to exchange data. They both "talk" on line A and "listen" on line B, a situation not conducive to successful communication. Similarly, two DCEs listen on line A and talk on line B. So, how does a null modem effect a solution to this problem? The simplest null-modem cables just cross lines A and B so that DTE 1 will be listening on the line that DTE 2 is talking on. Figure 3-6 illustrates the function of a null modem in interfacing two DCEs and two DTEs. Simple, yet elegant! There is an experiment dedicated to null modems.

Finally, to go back to our question about identifying the DCE and DTE in the configuration shown in Fig. 3-4, we can say that *usually* the CRT is the DTE and the computer is the DCE. If the computer does not already look like a DCE, then it is made to look like one by attaching a null modem to it.

Some additional words and definitions that are important to know before you begin to read the EIA standard appear below. Figure 3-7, which shows an RS-232-C interface circuit, illustrates the important concepts associated with these definitions.

INTERFACE POINT: The shared boundary in a physical connection where interface signals pass between equipment via electrical signals.

INTERFACE CIRCUIT: A circuit between the DTE and the DCE for the purpose of exchanging data, control, or timing signals. A Signal Ground circuit provides a common reference point for these signals.

DRIVER: The transmitter of a binary digital signal.

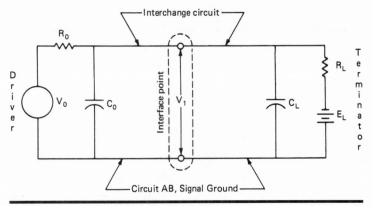

Fig. 3-7   RS-232-C interface circuit. (EIA.)

TERMINATOR: The receiver of a binary digital signal.

MARK: Equivalent to a positive logic 1.

SPACE: Equivalent to a positive logic 0.

SIMPLEX CHANNEL: A communications channel that is capable of operating in only one direction.

HALF-DUPLEX CHANNEL: A communications channel that is capable of operating in both directions but not simultaneously; i.e., the direction of transmission is reversible. Both directions of a half-duplex channel support the same range of data-transmission rates.

FULL-DUPLEX or DUPLEX CHANNEL: A communications channel that is capable of operating in both directions simultaneously. The range of data transmission rates supported is the same for both directions.

SYNCHRONOUS DATA-TRANSMISSION CHANNEL: A communication channel in which timing information is transferred between the DTE and the DCE.

NONSYNCHRONOUS DATA-TRANSMISSION CHANNEL: A communications channel in which no timing information is transferred between the DTE and DCE.

DATA SET: A device that uses incoming data to modulate an outgoing signal that is to be placed on switched or dedicated telephone network lines. The device also demodulates incoming signals from the telephone network and extracts the data that were used to modulate the signals at the originating end.

Scope

The RS-232-C standard applies to serial binary communications between DCEs and DTEs in which the data rates are in the range from 0 to 20,000 bps. Thus 19.2 kbps is the highest common data rate for which RS-232-C applies.

The standard places a rule-of-thumb limitation of 50 feet upon cable length. This limit can be exceeded if the specific environment is known and meets certain conditions that are detailed below.

The EIA specifically states that RS-232-C is not applicable to situations where electrical isolation between equipment on opposite sides of the interface is required. This is an important fact to keep in mind if you plan to employ the RS-232-C standard to interface your valuable computer to unproven, "home-grown" products of your own creative genius.

The fact that RS-232-C was intended to address local interfaces associated with long-distance data communications that involve the telephone network has not kept it from being applied to a wide variety of short-distance communications interfaces, such as computer to terminal, computer to printer, and even computer to disk. RS-232-C was really never intended to become the short-distance interface standard that it is today. This is just more evidence to support the saying that "Any standard is better than no standard." The null modem is no more than an elegant kludge to make such devices as computers and terminals (normally DTEs) look like modems (DCEs) so that RS-232-C will apply.

Electrical Signal Characteristics

The following summary of the RS-232-C electrical signal characteristics uses the labels shown in Fig. 3-7.

1. A signal on any pin of the RS-232-C connector has a *status* or *state* associated with it. The state is said to be one member of any of the following pairs of possible states:
   MARK/SPACE
   ON/OFF
   logic 0/logic 1

Table 3-1 shows the relationship between the above pairs and the signal voltage level. Note that RS-232-C employs negative logic, in that an ON condition is associated with logic 0, while an OFF condition is associated with logic 1. The signal voltage V_1 is measured with respect to a Signal Ground circuit that we discuss later. The range from -3 to $+3$ V is the transition region for which no signal state is given.

TABLE 3-1

Status	Signal Voltage	
	$-25\,V < V_1 < -3\,V$	$3\,V < V_1 < 25\,V$
Binary logic state	1	0
Signal condition	MARK	SPACE
Function	OFF	ON

2. To represent a logic 1 or MARK condition, a driver must apply a voltage between -5 and -15 V. To represent a logic 0 or SPACE condition, a driver must apply a voltage between $+5$ and $+15$ V.

Note that this rule, in conjunction with rule 1 above, implies that there is a 2-V noise margin in the standard. Figure 3-8 gives a graphical representation of the relationship between voltage level and signal status. Thus, a line driver, or signal source, *sends* a logic 0 by applying a voltage in the range $+5$ to $+15$ V. A line receiver, or signal destination, *senses* a logic 0 by looking for voltages in the range $+3$ to $+15$ V. The result is that there is allowance for a 2-V drop in a signal between its origination and destination. The allowance is similar for transmission of a logic 1.

Having now read about the voltage levels and their significance in an RS-232-C interface, it is reasonable to ask two specific questions:

1. Why depart from the TTL logic levels of 0 and $+5$ V?
2. Why pick -15 to -3 and $+3$ to $+15$ V?

First, the departure from TTL was motivated by the necessity to improve noise immunity and distance capabilities. Second, the voltage ranges of -15 to -3 and $+3$ to $+15$ were commonly available in most computer circuits at the time the RS-232-C standard was developed. Additionally, many off-the-shelf transistors are capable of working at these voltages, with the currents required to maintain them. The voltage ranges provide good noise immunity and allow operation at acceptable speeds (up to 20,000 bps). Moreover, MARKs and SPACEs are represented by opposite current flows and are differentiated by a minimum of 6 V. These attributes greatly enhance the reliability of data transfers.

3. The shunt capacitance C_L of the terminator side of an RS-232-C circuit must not exceed 2500 pF including the capacitance of the cable.

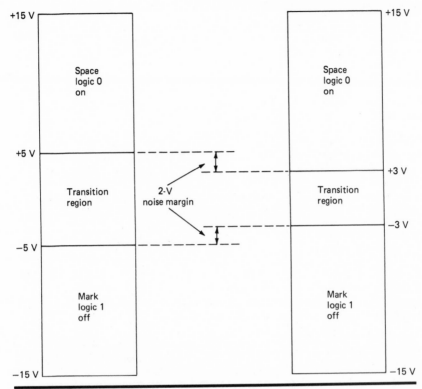

Fig. 3-8   RS-232-C interface electrical characteristics.

Note that this rule contributes significantly to the limitation of 50 ft that the standard places upon the length of an RS-232-C cable.

4. The open-circuit or no-load voltage V_o may not exceed 25 V. This just means that there are no voltages in an RS-232-C interface circuit that ever exceed 25 V.

5. An RS-232-C driver circuit must be able to withstand a short circuit to any other wire in the cable without sustaining damage to itself or any associated equipment. "Associated equipment" includes the terminal, modem, computer I/O port, and any other device that might be attached to the RS-232-C cable.

This rule means that if two pins are unintentionally shorted together, no damage should result.

Interface Mechanical Characteristics

The interface connector pin assignments in the RS-232-C standard are given in Table 3-2. Note that the RS-232-C standard makes no mention

TABLE 3-2   Interface connector pin assignments

Pin	Circuit	Description
1	AA	Protective Ground
2	BA	Transmitted Data
3	BB	Received Data
4	CA	Request to Send
5	CB	Clear to Send
6	CC	Data Set Ready
7	AB	Signal Ground
8	CF	Received Line Signal Detector
9/10	—	(Reserved for Data Set Testing)
11	—	Unassigned
12	SCF	Secondary Received Line Signal Detector
13	SCB	Secondary Clear to Send
14	SBA	Secondary Transmitted Data
15	DB	Transmit Signal Element Timing (DCE Source)
16	SBB	Secondary Received Data
17	DD	Receive Signal Element Timing
18	—	Unassigned
19	SCA	Secondary Request to Send
20	CD	Data Teminal Ready
21	CG	Signal Quality Detector
22	CE	Ring Indicator
23	CH/CI	Data Signal Rate Select (DTE/DCE Source)
24	DA	Transmit Signal Element Timing (DTE Source)
25	—	Unassigned

of the DB-25 (also called the D-type 25-pin connector) plugs and sockets that are typically associated with it. The DB-25 connector, which is almost always used in RS-232-C interfaces, is compatible with ISO 2113, a standard promulgated by the International Organization for Standardization (ISO). The document to obtain for more detailed information is:

ISO Draft International Standard 2110, "Data Communication: 25-Pin DTE/DCE Interface Connector and Pin Assignments" (Revision of ISO 2110-1972), February 1979.

Detailed mechanical drawings from this standard, of both the male (DB-25-P) and female (DB-25-S) connectors, are shown in Fig. 3-9. Note that a male connector is associated with DTEs; a female connector is associated with DCEs. Strictly speaking, an interface with 25 pins arranged in a figure 8 could pass for an RS-232-C mechanical connection. Don't lose any sleep over it, but you have been warned!

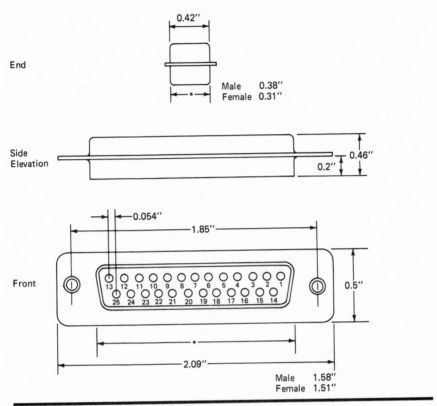

End

0.42"

Male   0.38"
Female 0.31"

Side Elevation

0.46"
0.2"

0.054"
1.85"

Front

13 12 11 10 9 8 7 6 5 4 3 2 1
25 24 23 22 21 20 19 18 17 16 15 14

0.5"

2.09"

Male   1.58"
Female 1.51"

Fig. 3-9   Mechanical drawings of DB-25 connectors.

Circuit Functional Characteristics

The circuits shown in Table 3-2 can be divided into five categories:

Ground or Common Return (A)

Data Circuits (B)

Control Circuits (C)

Timing Circuits (D)

Secondary Channel Circuits (S)

The letter in parentheses after each circuit category is the first in a two- or three-letter designation that is used throughout most of the literature that discusses RS-232-C signals. Table 3-3 groups the RS-232-C circuits by category, showing their two- or three-letter designation, their direction (to or from the DCE), and their cable pin assignment. Frankly, we

TABLE 3-3 **RS-232-C Interchange Circuits by Category**

Interchange circuit	Connector pin assignment	Description	Gnd	Data From DCE	Data To DCE	Control From DCE	Control To DCE	Timing From DCE	Timing To DCE
AA	1	Protective Ground	X						
AB	7	Signal Ground/Common Return	X						
BA	2	Transmitted Data			X				
BB	3	Received Data		X					
	4	Request to Send					X		
	5	Clear to Send				X			
	6	Data Set Ready				X			
	20	Data Terminal Ready					X		
	22	Ring Indicator				X			
	8	Received Line Signal Detector				X			
	21	Signal Quality Detector				X			
	23	Data Signal Rate Selector (DTE)					X		
	23	Data Signal Rate Selector (DCE)				X			
DA	24	Transmitter Signal Element Timing (DTE)							X
DB	15	Transmitter Signal Element Timing (DCE)						X	
DD	17	Receiver Signal Element Timing (DCE)						X	
SBA	14	Secondary Transmitted Data			X				
SBB	16	Secondary Received Data		X					
SCA	19	Secondary Request to Send					X		
SCB	13	Secondary Clear to Send				X			
SCF	12	Secondary Received Line Signal Detector				X			

find the two-letter circuit designator more confusing than enlightening. Hence, in the discussion below, while we include the shorthand code, it is only for purposes of completeness. In the future, we will identify RS-232-C circuits of interest by their function-derived name, e.g., Transmitted Data, rather than by code, e.g., circuit BA. The function of each of these circuits is described below.

Circuit AA: Protective Ground This conductor is electrically bonded to the machine or equipment frame. Since the Protective Ground circuit has no significance with respect to establishing a common signal ground reference, it is quite often left out. Its omission does not mean that a cable is noncompliant with the RS-232-C standard, since it is designated as optional.

Circuit AB: Signal Ground or Common Return This conductor provides the reference point relative to which all the other RS-232-C circuits, except Protective Ground (circuit AA), are measured. This is the *one* circuit that is absolutely *required*, no matter what the cable's application.

Circuit BA: Transmitted Data The signals on this circuit are transmitted from the DTE to the DCE. The DTE holds Transmitted Data (Circuit BA) at logic 1 (i.e., in a MARKing condition) at all times when no data are being transmitted.

In all systems compliant with RS-232-C the DTE must not transmit data unless a logic 0 (ON condition) is present on each of the following circuits, *where implemented:*

1. Request to Send (circuit CA)
2. Clear to Send (circuit CB)
3. Data Set Ready (circuit CC)
4. Data Terminal Ready (circuit CD)

We will discuss the several ways in which a system can reach such a state in the next section.

For a microcomputer system with a local terminal connected to it via an RS-232-C interface, if the terminal is acting as the DTE (as is usually the case), then the computer must be acting as a DCE. Thus, the terminal (DTE) talks on the Transmitted Data line, and the computer (DCE) listens on this circuit. The term *transmit* is used relative to the DTE, not to the DCE. Note that there is a potential for trouble if the DCE attempts to act like a DTE and transmit on this circuit. This is precisely the situation that calls for a null modem.

Another important fact is that quite often some of the circuits that must be ON before transmission can occur are not implemented. The lack of implementation can be due to the terminal, to the computer I/O port, or to the cable that attaches the terminal to the computer. Quite often, it is the cable. If you are having trouble achieving communications over an RS-232-C interface, the probability is quite high that your problem stems from one of the following: (1) one of the above signals is OFF or (2) your terminal is transmitting on the same line as the computer is (hence your terminal is not listening to the computer and vice versa). These and other common pitfalls are explored in the experiments at the end of the chapter.

Circuit BB: Received Data The signals on this circuit are transmitted from the DCE to the DTE. The circuit is held at logic 1 (MARKing) during intervals between data transmission and at all times when no data are being transmitted.

On a half-duplex channel, Received Data (circuit BB) is held in an OFF condition when Request to Send (circuit CA) is in the ON condition. Also, Received Data is held OFF for a brief interval after an ON to OFF transition of the Request to Send line to allow for completion of transmission.

Circuit CA: Request to Send This circuit carries a request to transmit from the DTE to the DCE. Figures 3-9 and 3-10 illustrate how the Request to Send (RTS) signal works with another signal called Clear to Send (CTS or circuit CB) to coordinate data transmission between the DTE and DCE.

For simplex and full-duplex channels, a logic 0 on the Request to Send line keeps the local DCE in transmit mode. In this context, a DCE in transmit mode means that the DCE will accept data from the DTE and pass it on (in unchanged form) to the communications link to which it is attached. For example, if the DCE is a modem, then *transmit mode* implies that the modem will transmit the data it receives from the DTE to the telephone network. Conversely, for simplex and half-duplex communication channels, if the Request to Send signal is at logic 1 (i.e., OFF), the local DCE is maintained in a nontransmit mode. *Nontransmit mode* means that the DCE will not pass data it has received from the DTE on to the communications network.

For half-duplex communication channels, the ON condition maintains the local DCE in the transmit mode, and the OFF condition maintains the local DCE in the receive mode. Of course, *receive mode* here means that the DCE will accept data from the communications network and pass it on (in unchanged condition) to its local DTE.

Referring to Fig. 3-10, let us look at the significance of transitions (ON to OFF and OFF to ON) on the Request to Send line. A transition of Request to Send from OFF to ON triggers the local DCE to enter trans-

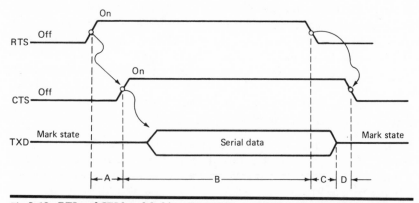

Fig. 3-10 RTS and CTS handshaking: (A) DTE asserts RTS, indicating that it wishes to transmit data; DCE is activated to set up communications; DCE asserts CTS, indicating that DTE should commence transmission; TXD is in idle marking condition. (B) Data transmission occurs on the transmitted data circuit; DTE deactivates RTS, indicating that it no longer wishes to transmit. (C) DCE completes transmission of last data; TXD returns to idle marking condition. (D) DCE readies itself to respond to next assertion of RTS; DCE indicates that it is ready for next RTS by turning off CTS.

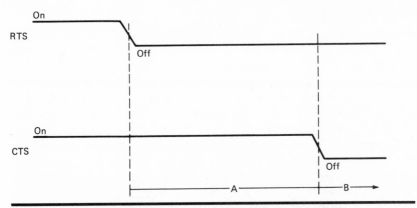

Fig. 3-11 *When RTS can be asserted: (A) During this interval, the DTE may not reassert RTS; presumably the DCE is cleaning up from the last transmission and will indicate its readiness for subsequent transmission by turning off CTS. (B) Now that CTS is off, the DTE is free to assert RTS.*

mit mode and perform such actions as are necessary to set up communications. These "setup" activities range from dialing up a remote host using an autodialing unit to no action whatsoever. Most commonly, in microcomputer systems, you will find the Request to Send line jumpered directly to the Clear to Send line. The net effect of this arrangement is that as soon as the DTE asserts Request to Send, it immediately receives back this signal in the form of an asserted Clear to Send signal. (This will be investigated in more detail in an experiment.) Once the setup actions (whatever they are) are successfully completed, the DCE then turns ON Clear to Send (circuit CB), thereby indicating to the local DTE that data may be transferred across the interface point on the Transmitted Data circuit.

Note that a transition of the RTS line from the ON to the OFF condition instructs the local DCE to complete transmission of all data that have previously crossed the interface point on the Transmitted Data circuit and then assume a nontransmit mode (full-duplex or simplex) or receive mode (half-duplex). The local DCE responds to this signal by turning OFF Clear to Send when it is prepared to respond to a subsequent ON condition of the RTS circuit.

As Fig. 3-11 shows, once Request to Send has been turned OFF, it may not be turned ON again until Clear to Send has been turned OFF by the local DCE. Note that this rule prevents overrun errors in which the DTE resumes transmission before the DCE has completed the previous transmission.

Depending upon the particular DTE and DCE involved, the Request to Send–Clear to Send handshaking just described can occur on a character-by-character basis, on a block-by-block basis, or not at all. A higher-level protocol (than the physical-level protocol) determines what constitutes a character or block. If 10 bits constitutes a character, then character-by-character handshaking would require the DTE to assert the Request to Send line and receive back a Clear to Send from the DCE for each ten bits sent. For block-by-block handshaking the DTE sends a special end-of-transmission character and turns off Request to Send at the end of each block transfer. In response to this, the DCE turns OFF Clear to Send once the end-of-transmission character has passed through it to the communications network. In the case where no hand-shaking occurs, Request to Send, Clear to Send, or both are held constantly high or looped-back by circuitry in the involved equipment. You will employ a loop-back technique in the experiments.

Where implemented, Data Terminal Ready and Data Set Ready must have been asserted, Request to Send must have been turned ON, and acknowledgment must have been received in the form of an asserted Clear to Send line before the DTE can transmit data on the Transmitted Data line. It is permissible to turn ON Request to Send at any time when Clear to Send is OFF, regardless of the condition of any other interface circuit.

Note that the Request to Send control signal coordinates the actions of the local DCE and DTE. Since telephone lines, microwave links, and satellite links may potentially separate the local DCE and remote DCE, this signal is *not* available to any remote equipment. Just as importantly, it does *not* imply anything about the status of any remote equipment.

Again, the absence of Request to Send is sometimes the problem in achieving successful RS-232-C interfaces. This is especially true in microcomputer systems that utilize USARTs in their serial I/O port circuits. We will explore this more in the experiments at the end of the chapter.

Circuit CB: Clear to Send   This is a control signal that is transmitted from the DCE to the DTE to indicate that the DCE is ready to receive data from the DTE on the Transmitted Data circuit. When this signal is ON and the signals Request to Send, Data Set Ready, and Data Terminal Ready are all on, this constitutes an indication to the DTE that data transmitted by it will be communicated by the DCE to the communications channel. When Clear to Send is OFF, it indicates that the DCE is not ready so the DTE should not attempt to transmit data.

Clear to Send is turned on in response to simultaneous ON conditions of Request to Send (circuit CA) and Data Set Ready (circuit CC). If Request to Send is not implemented, then it should be assumed to be constantly in the ON condition, and the Clear to Send circuit should respond according to activation and deactivation of the Data Set Ready signal.

The following circuits are all required in RS-232-C compliant interfaces between DTEs and DCEs that involve the public switched telephone network:

Circuit CC   Data Set Ready

Circuit CD   Data Terminal Ready

Circuit CE   Ring Indicator

Circuit CF   Received Line Signal Detector

Since their functions are quite specific to the telephone network, they are often omitted from the purely local, short-distance brand of RS-232-C interfaces so common in microcomputer systems. Some of the most interesting applications of these signals involve automatic dialing equipment and automatic answering equipment.

The major significance of the two signals, Data Set Ready and Data Terminal Ready, is that they constitute equipment readiness indicators. That is, if Data Set Ready is ON, its associated DCE is ready to pass data from the DTE to the network. Similarly, if Data Terminal Ready is ON, then the DTE is ready to transmit data to the DCE on the Transmitted Data circuit. As mentioned earlier in our discussion of the Transmitted Data circuit, both Data Set Ready and Data Terminal Ready must be in an ON condition prior to data transmission in an interface compliant with RS-232-C. Since these signals are not implemented in most microcomputer systems, data transmission, in this case, is predicated upon activation of Request to Send and Clear to Send.

Circuit CC: Data Set Ready   The direction of this control signal is from the DCE to the DTE. It indicates the status of the local data set. If the Data Set Ready signal is ON, it means that the DCE is connected to the communication channel. In automatic calling situations, this means that the DCE has dialed the number, completed call establishment, and is in data-transmission (as opposed to voice-transmission) mode.

Circuit CD: Data Terminal Ready   The direction of this control signal is from the DTE to the DCE. Data Terminal Ready must be ON before the DCE can turn ON Data Set Ready, indicating that it has been connected to the communications channel. Once the DCE is connected and data are being transmitted, a transition of Data Terminal Ready from the ON to the OFF condition causes the DCE to be removed from the communication channel.

Essentially, Data Terminal Ready and Data Set Ready implement a more static version of the Clear to Send–Request to Send protocol. By more static we mean that where the Clear to Send–Request to Send protocol addresses channel readiness, the Data Terminal Ready–Data Set Ready protocol addresses equipment readiness. Equipment readiness is normally a much less volatile condition than channel readiness.

Circuit CE: Ring Indicator The direction of this control signal is from the DCE to the DTE. When this signal is ON, it indicates that the DCE is receiving a ringing signal. The signal is maintained in the OFF condition between "rings" and at all other times when the DCE is not receiving a ringing signal. The major application of this RS-232-C control signal is in configurations with automatic answer modems.

Circuit CF: Received Line Signal Detector The DCE sends an ON condition to the DTE on this circuit when it is receiving a carrier signal that meets its suitability criteria from the remote DCE. These criteria are determined by the manufacturer of the DCE. A widely used alternative name for this signal is *Data Carrier Detect (DCD)*. On modems, this line is usually connected to an LED indicator labeled *Carrier*.

There are many terminals on the market that require this signal to be constantly high before they will either send or receive data. Such a terminal is the Texas Instruments TI Silent 700 series, which is equipped with an acoustic-coupled internal modem and an RS-232-C port for communication via an external modem. The Received Line Signal Detector is asserted when the carrier from the remote answering modem is sensed if the internal modem is used. If, on the other hand, one wishes to interface a Silent 700 to a local microcomputer (or similar equipment), then an asserted DCD must still be present. Typically, one accomplishes this by tying the signal to some other pin on the RS-232-C connector that is always asserted, such as Data Terminal Ready. This technique, known as *jumpering*, employs a small jumper or wire to form an electrical connection between two pins within the connector itself. In this way, one can tamper with the RS-232-C control logic via modifications to a relatively inexpensive cable instead of the DCE or DTE, which are typically more costly items.

The remaining circuits are:

Circuit CG	Signal Quality Detector
Circuit CH	Data Signal Rate Selector (DTE Source)
Circuit CI	Data Signal Rate Selector (DCE Source)
Circuit DA	Transmit Signal Timing (DTE Source)
Circuit DB	Transmit Signal Timing (DCE Source)
Circuit DD	Receiver Signal Timing (DCE Source)

Circuit SBA   Secondary Transmitted Data

Circuit SBB   Secondary Received Data

Circuit SCA   Secondary Request to Send

Circuit SCB   Secondary Clear to Send

Circuit SCF   Secondary Receive Signal Detector

These circuits are of no significance and are left unconnected in the vast majority of microcomputer configurations. For this reason, we will not discuss them at all here. The interested reader is encouraged to read the relevant sections of the RS-232-C standard.

COMMON CONFIGURATIONS

There is a great deal of latitude in the RS-232-C standard regarding what constitutes a compliant cable-connector implementation. First, as we mentioned earlier, the connector itself is left completely unspecified. There is an appendix to the standard in which the EIA states that commercial connectors which meet Military Specification MIL-C-24308 (MS-18275) will perform satisfactorily in an RS-232-C interface. However, shortly after this recommendation, the EIA states in no uncertain terms that the MIL standard should *not* be considered as part of the RS-232-C standard.

Another area in which there is great latitude pertains to the signals that appear in the RS-232-C cable. The standard defines 21 circuits. Which ones are required and which ones are optional? This is an extremely difficult question to answer because the standard can apply to so many different types of communications configurations. These configurations range in complexity from a simple local terminal interface with a microcomputer to a multiplexed, synchronous, dedicated line that is shared by a cluster of remote terminals and is equipped with automatic dialing units. Wisely, the EIA has developed a standard cable for several distinct situations. Seven of their sample configurations are applicable to microcomputer systems:

Transmit only

Transmit only with RTS

Receive only

Half Duplex

Full Duplex

Full Duplex with RTS

Special

TABLE 3-4 RS-232-C Standard Configurations

	RS-232-C interchange circuit	Transmit only	Transmit only with RTS	Receive only	Half duplex	Full duplex	Full duplex with RTS	Special
1	Protective Ground	—	—	—	—	—	—	o
7	Signal Ground	X	X	X	X	X	X	X
2	Transmitted Data	X	X			X	X	o
3	Received Data			X	X	X	X	o
4	Request to Send	X	X		X	X	X	o
5	Clear to Send	X	X		X	X	X	o
6	Data Set Ready	S	S	X	X	X	X	o
20	Data Terminal Ready	S	S	S	S	S	S	o
22	Ring Indicator			S	S	S	S	o
8	Received Line Signal Detector			X	X	X	X	o

X = required for any configuration.
S = required for using public switched telephone network.
o = specified by cable designer.

Table 3-4 shows each of the above interface types and the RS-232-C signals that must be present in a standard cable. Note that we have omitted the signal element timing circuits that would be required for synchronous transmission, as synchronous communications occur so rarely in microcomputer systems.

As Table 3-4 shows, only *one* circuit is an *absolute requirement* for any RS-232-C cable, namely Signal Ground on pin 7. Every other signal may or may not be present, depending upon the intended application of the cable. The designation "special" gives license to cable designers and manufacturers to include any signals at all, and, as long as Signal Ground is included on pin 7, they can justifiably say that the cable is "RS-232-C compliant."

While all the standard cables shown in Table 3-4 are applicable to microcomputer systems, the vast majority of microsystems utilize some variant of the full-duplex RS-232-C cable. Certainly many microcomputer systems incorporate such one-way devices as receive-only printers, transmit-only joysticks, and similar devices. But most of the time, the RS-232-C cables used are configured to handle the most common situation, namely full-duplex, two-way communication. In this way, the cable normally used for a receive-only printer can be moved to different devices as needed, with no ill effects. The flexibility realized by having several interchangeable cables is almost always justification for including a few extra circuits.

The classical application of full-duplex communications in microcomputer systems is for send/receive terminal communications where characters are transmitted from a keyboard to a microcomputer and are subsequently echoed back to the display screen or print mechanism. In this situation data are simultaneously traveling in both directions between the DTE (keyboard and display device) and the DCE (computer serial I/O port). A typical full-duplex communications interface appears in Fig. 3-12. Table 3-4 lists two full-duplex standard cables, one with and one without the Request to Send line implemented. Thus, as far as the EIA is concerned, the presence of the Request to Send line in a full-duplex cable is optional. As you may well experience, if your computer I/O ports use USART integrated circuit serial I/O drivers, this line is *not* optional for you. As a conservative, safe strategy, when we make up cables for ourselves, we always include the Request to Send circuit, either as a conductor in the cable or as a jumper to Clear to Send. The first several experiments at the end of this chapter show you detailed schematics of RS-232-C cables that work in most microsystem applications, as well as some "foolproof" procedures for determining your exact, system-specific cable requirements. The point that we wish to emphasize here is that, while the EIA's standard full-duplex cable

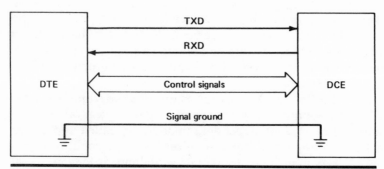

Fig. 3-12 A full-duplex configuration.

exists and has broad applicability to communications usually involving modems and the telephone network, as microsystem users you must go beyond the standard so that you can put it to uses for which it was never intended!

The first step in going beyond a standard is to understand what you propose to extend. Note that in this context, by *extend* we mean that you will be designing your own custom RS-232-C cables, i.e., cables that would fall under the special category in the EIA's set of standard cables. The price of this freedom from the bondage of the EIA's nonspecial, prespecified cables is *knowledge*. If you are aware of the function of each applicable RS-232-C interface circuit, you can make an informed decision to include or exclude it in your special cable. Thus, the goal of the following section is to explore in detail the function of each circuit in the RS-232-C cable for full-duplex communications.

RS-232-C STANDARD FULL-DUPLEX CABLE

The signals present in a standard full-duplex RS-232-C cable are:

Signal Ground

Transmitted Data

Received Data

Request to Send

Clear to Send

Data Set Ready

Received Line Signal Detector

If the DCE is a modem that is communicating over the switched telephone network, i.e., via nondedicated, public routes, then two more signals must be added to the above group:

Data Terminal Ready

Ring Indicator

To understand the function of each of the above mandatory signals, consider the configuration shown in Fig. 3-13. In order for the two computer systems in Fig. 3-12 to exchange data, several events must take place. Each event causes the configuration to undergo some sort of transition. For example, if the telephone is ringing and the call answering station "answers the phone," the answering DCE makes a transition from a disconnected state to a connected state (with respect to the telephone network). The goal is to put together a set of "correct" events, in a correct sequence, to cause the equipment to pass from the idle, not-communicating state to a target data-exchange or communicating state and then finally back to the original idle state. There are several event sets and sequences that will accomplish this. The EIA has grouped

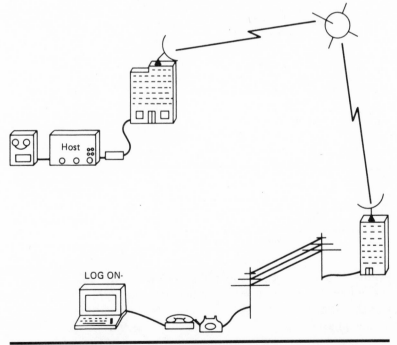

Fig. 3-13   Configuration for using the public switched telephone network.

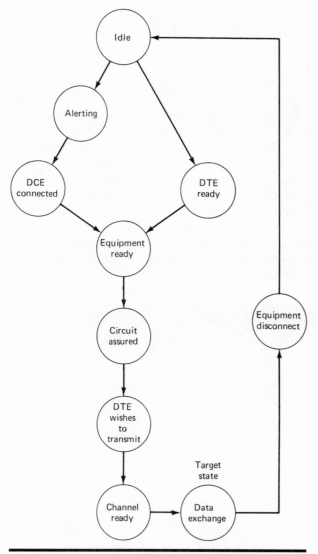

Fig. 3-14   Summary: RS-232-C full-duplex control logic.

these events and transitions into phases between the idle and data-exchange states that the equipment must pass through. These phases are summarized in Fig. 3-14.

The major phases are:

Alerting

Equipment Readiness

Circuit Assurance

Channel Readiness

In the Alerting phase, the call-originating station, among other things, dials the telephone number of the remote call-answering station. When the remote telephone begins to ring, the Ring Indicator signal of the answering DCE makes an OFF to ON transition. When the answering DCE "answers the phone," the Alerting phase ends. Overlapping in time with the Alerting phase is the Equipment Readiness phase. The RS-232-C signals involved in this phase are Data Set Ready and Data Terminal Ready. When events associated with turning ON both of these signals have occurred, the Equipment Readiness phase is complete. The Circuit Assurance phase consists of events associated with turning ON the Received Line Signal Detector signals at both communicating stations. Once the circuit is assured, the Channel Readiness phase uses Request to Send and Clear to Send handshaking to arrive at the target state of active data exchange. Several events can cause a transition out of the data-exchange state. All these events result in the equipment being disconnected from the telephone network and ultimately reverting back to the original idle state.

In the above paragraph, we were not very specific about particular events and their relative occurrence in time. For this, we will use a mechanism called *state-transition diagrams*. These state-transition diagrams graphically illustrate the function of each full-duplex RS-232-C signal in two communicating DCE-DTE stations. Essentially, a change in logic state of each RS-232-C signal corresponds to an event whose occurrence is associated with a transition of the DCE-DTE station from one state to another. The purpose of the state-transition diagrams is to give detailed information describing the states that a station can assume and the effects of various events upon a station in a particular state. Once you have this information, you can use the RS-232-C signals to fully control the state of a DCE-DTE station. Figures 3-15 and 3-16 give state-transition diagrams for equipment that originates and equipment that answers a call, respectively.

The next two sections walk through the state-transition logic that governs the behavior of call-originating and call-answering equipment using a standard full-duplex RS-232-C interface.

State-Transition Logic: Call-Origination Equipment

In computer science (as well as other disciplines), a very useful method for providing a clear and unambiguous description of rather complicated event sequences is to use state-transition diagrams. In Fig. 3-15,

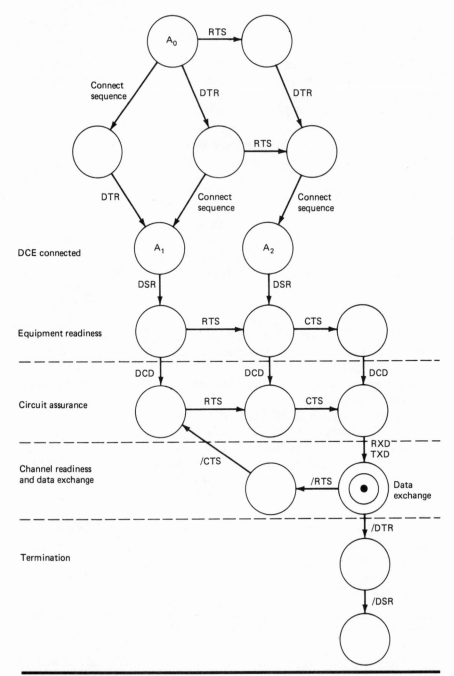

Fig. 3-15 Full-duplex communications state-transition diagram: call-origination equipment.

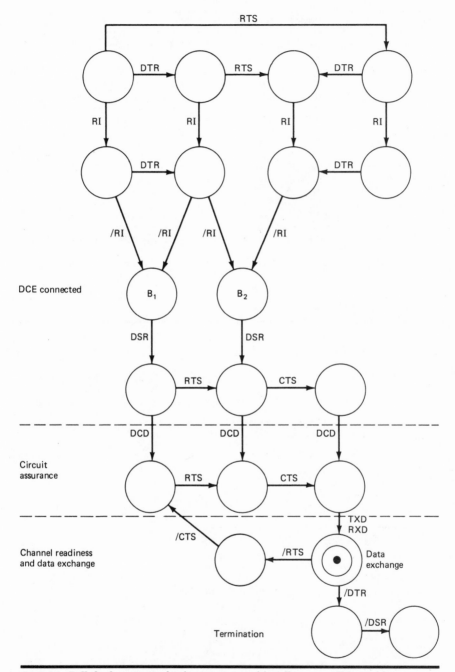

Fig. 3-16   Full-duplex communications state-transition diagram: call-answering equipment.

TABLE 3-5 State space for full-duplex RS-232-C communications

GROUP A: ALERTING AND EQUIPMENT READINESS				GROUP B: CIRCUIT ASSURANCE AND CHANNEL READINESS			
Component identifier	Data set ready	Data terminal ready	Ring indicator	Component identifier	Request to send	Clear to send	Received line signal detector
0	OFF	OFF	OFF	0	OFF	OFF	OFF
1	OFF	OFF	ON	1	OFF	OFF	ON
2	OFF	ON	OFF	2	OFF	ON	OFF
3	OFF	ON	ON	3	OFF	ON	ON
4	ON	OFF	OFF	4	ON	OFF	OFF
5	ON	OFF	ON	5	ON	OFF	ON
6	ON	ON	OFF	6	ON	ON	OFF
7	ON	ON	ON	7	ON	ON	ON
X	?	?	?	X	?	?	?

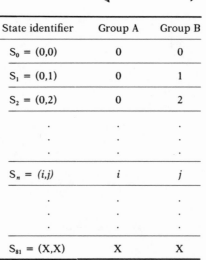

State identifier	Group A	Group B
$S_0 = (0,0)$	0	0
$S_1 = (0,1)$	0	1
$S_2 = (0,2)$	0	2
. . .	. . .	. . .
$S_n = (i,j)$	i	j
. . .	. . .	. . .
$S_{81} = (X,X)$	X	X

each circle represents a state that the equipment can assume. A special state that is desirable to attain, called a *target state*, is drawn with two concentric circles. The arrows connecting the states represent *events* that take the equipment *from* the state at the tail of the arrow *to* the state at the head of the arrow. A significant strength of this method of representing equipment behavior is that it graphically shows how the equipment is influenced both by being in a certain state and by the occurrence of a certain event. Note that the effect of a particular event will vary depending upon the state of the equipment when the event occurs. In the context of a state-transition diagram, the effects of events are *transitions*.

A crucial prerequisite to expressing equipment behavior in terms of states, events, and transitions is to fully define the following:

1. The legal states that the subject equipment may assume
2. The set of events that may occur
3. The transition caused by each event for equipment in each possible state

Table 3-5 presents a formal definition of all the possible states that call-origination equipment can assume, relative to a full-duplex RS-232-C standard interface. Notice that all we have really done is enumerate the possible combinations of logic levels for the six control signals in the RS-232-C full-duplex cable. Specifically, what we did was divide the six control signals into two groups. Group A consists of the three signals associated with the Alerting phase and the Equipment Readiness phase, namely Ring Indicator, Data Set Ready, and Data Terminal Ready. Group B consists of the signals associated with the Circuit Assurance phase and the Channel Readiness phase, namely Request to Send, Clear to Send, and Received Line Signal Detector. Since groups A and B are both comprised of three signals, each group can define 2 cubed (or 8) plus 1 (X = don't care), or 9, possible states. Note that the "don't care," or X, condition just means that the three signals in the subject group (A or B) may assume any status without any impact upon the state of the interface as a whole. Combining groups A and B gives a comprehensive description of the status of a DCE-DTE station. Therefore, a *state* is given by a pair of numbers: the first gives the conditions of the group A signals, and the second gives the conditions of the group B signals. For example, the state (6,7) describes a DCE-DTE station whose RS-232-C interface signals are in the following condition:

Data Set Ready:	ON
Data Terminal Ready:	ON
Ring Indicator:	OFF

Request to Send:	ON
Clear to Send:	ON
Received Line Signal Detector:	ON

Once you have stepped through the state-transition logic for the two types of stations (origination and answering) you will see that equipment in state (6,7) has reached the target state and has completed all the prerequisites for transmitting data to the remote station.

As another example, consider the state given by the pair (0,X). In this state, Ring Indicator, Data Terminal Ready, and Data Set Ready are all OFF. Here, the 0 indicates that Ring Indicator, Data Terminal Ready, and Data Set Ready are all OFF. The X indicates that since all the group A signals are OFF, and the system is idle, the status of the group B signals is of no interest because the status of the interface as a whole will remain the same. In particular, (0,0) and (0,1) and. . .and (0,7) are all idle states, and, for our purposes, no distinction between them is either necessary or desirable.

In the above enumeration of possible states we did not include logic levels on some of the signals present in a full-duplex standard RS-232-C cable. It is natural to ask why. The answer to this question becomes clear when you look at what those signals are:

- *Signal Ground:* Clearly this signal, once its presence is insured, has no influence on the equipment's readiness for exchanging data.
- *Transmitted Data:* This signal is important when the equipment is actively sending data, i.e., when the target state of active data exchange has been reached. The transmitted data line has no significant role in the event sequence leading either in to or out of the target state.
- *Received Data:* The role of this signal is similar to that of Transmitted Data. It is obviously very significant once the target state has been reached, but it has no role in achieving that state.

Table 3-6 shows the set of possible events that will cause transitions among the states shown in Table 3-5. The event that initiates the series of transitions ultimately leading to the target (data-exchange) state is the *connect sequence* associated with physically connecting the DCE to the telephone network. The other events in the table all relate to logic state changes in the six control signals in the RS-232-C cable. The notational convention used is as follows:

XXX represents an OFF to ON or logic 1 to logic 0 transition of the control signal whose tag is XXX.

/XXX represents an ON to OFF or logic 0 to logic 1 transition of the control signal whose tag is XXX.

TABLE 3-6   Event space for full-duplex RS-232-C communications

Event identifier	Description
Connect sequence	A series of events initiated by RI for call-answering equipment and initiated by Decision to Initiate a Call for call-origination equipment. See Figs. 3-15, 3-16.
RI	An OFF to ON transition of the Ring Indicator circuit
/RI	An ON to OFF transition of the Ring Indicator circuit
DTR	An OFF to ON transition of the Data Terminal Ready circuit
/DTR	An ON to OFF transition of the Data Terminal Ready circuit
DSR	An OFF to ON transition of the Data Set Ready circuit
/DSR	An ON to OFF transition of the Data Set Ready circuit
RTS	An OFF to ON transition of the Request to Send circuit
/RTS	An ON to OFF transition of the Request to Send circuit
DCD	An OFF to ON transition of the Received Line Signal Detector circuit; DCD stands for Data Carrier Detect
/DCD	An ON to OFF transition of the Received Line Signal Detector circuit
CTS	An OFF to ON transition of the Clear to Send circuit
/CTS	An ON to OFF transition of the Clear to Send circuit

The signal tags and event identifiers associated with each signal (that is, XXX and /XXX) are given in Table 3-6.

Now, referring back to Fig. 3-15, we can interpret the state-transition behavior of call-origination equipment in a full-duplex RS-232-C interface. The first important transition occurs from the idle, disconnected state to the state of being connected to the telephone network. The event associated with this transition is the *connect sequence*. It is possible to dissect the *connect-sequence* event into smaller subevents. This is done, for call-origination equipment, in Fig. 3-17. The method by which the transitions of Fig. 3-17 are accomplished can be manual or automatic. The initial triggering event is the decision to originate a call. Once this decision is made, for manual call origination, the telephone handset is taken off-hook, a dial tone is received, the telephone number of the remote station is dialed, the remote station answers the call, and the local-to-remote DCE connect sequence is complete.

Now that the local DCE is connected to the telephone network, what

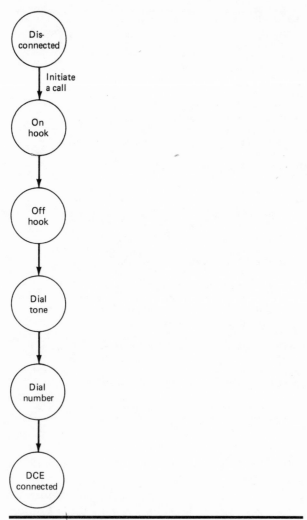

Fig. 3-17 Call-origination connect sequence.

about the local DTE? It has to move from an idle to a ready condition
and notify its local DCE when it has. For the DTE, a ready condition is
defined as Data Terminal Ready in an ON condition. So in the context
of the state-transition diagram, one event must occur, namely DTR.
Quite often, DTEs are configured to assert their Data Terminal Ready
line when power is applied, thus equating the state of being ready with
that of being turned on. As you can easily see from Fig. 3-15, the connect-sequence and Data Terminal Ready events must both occur before
the DCE turns ON the Data Set Ready signal, but the order in which

these two events takes place is not significant. This is illustrated graphically by showing two paths in to the state labeled A_1. The only difference between states A_1 and A_2 is whether or not the Request to Send event has occurred. In particular, state A_1 is achieved by either of the following two event sequences:

Connect sequence → DTR
DTR → Connect sequence

State A_2 is achieved by any one of the following six event sequences:

DTR → RTS → Connect sequence
DTR → Connect sequence → RTS
RTS → DTR → Connect sequence
RTS → Connect sequence → DTR
Connect sequence → DTR → RTS
Connect sequence → RTS → DTR

To keep our state-transition diagram simple, we have not explicitly represented all the above possibilities.

The above set of six possible paths to state A_2 gives you some idea of the proliferation of legal event sequences that results when a signal such as Request to Send is not constrained to rigid sequencing requirements. Looking at the entire state-transition diagram, you can see that the Request to Send event may occur before or after Data Terminal Ready, Data Set Ready, and Data Carrier Detect without any significant effect upon ultimately reaching the data-exchange state. A formal, complete exposition of DCE-DTE behavior should include analysis of all possible event sequences, in addition to the comparatively straightforward sequences that we have just mentioned. For example, what happens if /DSR happens just before DTR, just after DTR, or just after RTS? What happens if other events, such as RI, DCD, or CTS, occur? As you can see, the combinations and permutations of event sequences soon reach a point beyond which their enumeration and analysis produce diminishing returns. We leave such comprehensive analysis to the EIA, since they, as the authors of the standard, are obliged to follow all these possibilities to their ultimate resolution. Since most of these combinations are impossible, illogical, or, at best, unenlightening, we shall press on with our consideration of the "interesting" subset of event sequences.

We have progressed to state A_1 or A_2, depending upon whether or not we wish to assume that the DTE has asserted its Request to Send line. The next important event, in either case, is DSR. The DCE turns ON its Data Set Ready signal only when the connection sequence is complete

AND when the DTE has indicated its readiness via the DTR event. The DSR event marks the completion of the Equipment Readiness phase. The next phase to complete is that of Circuit Assurance.

Given that DSR and DTR have occurred, the DCE places its transmit-data carrier on the telephone connection between it and the remote DCE. Ideally, the remote DCE has acted likewise, and before long, the local DCE will detect the remote DCE's transmit carrier. When this happens, the local DCE turns ON its Received Line Signal Detector signal; i.e., the DCD event occurs. Since we are assuming full-duplex communications, DCD events at both the local and remote communicating stations (with no intervening /DCD events anywhere) constitute *circuit assurance*. Circuit assurance means that both DCEs are listening and can hear each other, a serendipitous state!

With the Equipment Readiness and Circuit Assurance phases now behind us, there remains only the Request to Send–Clear to Send handshaking separating us from the target data-transmission state. The DTE must only assert Request to Send to receive back a "go-ahead" signal from the DCE in the form of Clear to Send. Note that the Request to Send and Clear to Send events may then be followed by the sequence /RTS → /CTS → RTS → CTS on a block-by-block or character-by-character basis as data are transmitted and received by the local station. In particular, when the local DTE is transmitting, both the local Request to Send and Clear to Send events have occurred. When the local DTE decides to stop transmitting, it generates a /RTS event, which, in turn, causes the DCE to complete transmission of the data it has received and then to generate /CTS. Meanwhile, the remote DTE may have been sending data in the form of echoed characters back to the local DTE, or it (the remote DTE) may have some data to send to the local DTE in response to what it just received. The point that we wish to emphasize here is that turning OFF its Request to Send line in no way disables a DTE from receiving data.

This last point brings up a fairly important concept that has great significance at all levels of data communications. Of the (at least) two participants in a data exchange, receiver and transmitter, the target of all control logic is almost always the *transmitter*. Notice that RS-232-C does not have a "Request to Receive" line! Notice that a modem places a transmit carrier on the telephone network, not a "receive carrier." The reason for this emphasis upon the transmitter in RS-232-C control logic (as well as other communications protocols) is that a transmitter is active and a receiver is passive. Hence, by controlling the active member of a communicating pair, one achieves control of both. Additionally, transmission consumes communications link resources or bandwidth, while reception does not. In busing architectures, it is common to have a single transmitter broadcasting messages to several receivers. There can be only one transmitter talking at a time. There is no theoretical limit to the number of receivers that can be listening.

Therefore, when a DTE-DCE station is acting as a transmitter, it must follow the Request to Send–Clear to Send protocol. When the station is in a passive role as receiver, Request to Send and Clear to Send have no significance as events. In its role as receiver, the DCE passes all data from the telephone network side to the DTE side of the interface, no matter what the status of either Request to Send or Clear to Send.

Once the target state is achieved, data exchange continues until some event occurs to terminate it. The normal terminating event is /DTR. As soon as /DTR happens, the DCE turns OFF the Data Set Ready line (/DSR) and disconnects itself from the telephone network. This event sequence is shown in Fig. 3-14.

One interesting abnormal termination sequence (not shown in Fig. 3-14) that occurs quite often to users of remote time-sharing systems is triggered when the user removes the telephone handset from the acoustic coupler of the local modem. Some automatic answer modems that are used by time-sharing services are configured to turn OFF the Clear to Send line if the Received Line Signal Detector goes OFF. This means that if the handset is removed from the user's acoustic coupler, the remote modem no longer senses a carrier from the user's DCE. The term *carrier* will be discussed in detail later in the chapter on modems. For now, it is sufficient to think of it as a signal that a modem either modulates or demodulates for the purpose of sending data over the telephone network. Hence, in this sense, the signal "carries" the data from a transmitting DCE to a receiving DCE. When the remote host computer's DCE loses the user DCE's carrier, the result is that, in the remote host station, /DCD occurs, which triggers /CTS. /CTS causes the host DTE to stop data transmission to its local DCE, and hence to the user. If the telephone network connection is not dropped, it is actually possible for the user to put the handset back into the acoustic coupler and resume transmission with very minimal loss of data. On the other hand, most host automatic-answer modems are designed to drop the telephone connection if the remote DCE's carrier disappears from the line. The basic philosophy governing the host's behavior in each of the above scenarios can be simply put: "If I can't hear the user, the user probably can't hear me, so I'll quit." The meaning of quit can be either "stop transmitting data" or "cut the telephone connection." If the latter interpretation is the one in effect, then there is no possibility of alternating voice and data communications. It is often quite nice to have that option, so several modems of the nonautomatic variety do not "hang up" when /DCD occurs.

State-Transition Logic: Call-Answering Equipment

Figure 3-16 is the state-transition diagram for a DTE-DCE station that answers (as opposed to originates) incoming calls from any of several possible remote stations. As was the case with call-origination equipment, we must characterize the possible system states and the set of events that cause transitions between states. Tables 3-5 and 3-6 give the states and events that characterize call-answering equipment as well as call-origination equipment. As you will see below, the major difference between call-origination and call-answering stations occurs during the early phases, i.e., the Alerting and Equipment Readiness phases.

As Fig. 3-16 shows, the initial triggering event that causes an answering station to move from the idle, disconnected state to the connected state occurs when the telephone starts to ring. Figure 3-18 shows the breakdown of the connect-sequence event for call-answering equipment. The RS-232-C control signal, Ring Indicator, is ON when the telephone is ringing, and OFF between rings. When the Ring Indicator is ON, the DCE either automatically or with manual assistance takes the telephone off-hook and connects the DCE to the telephone network.

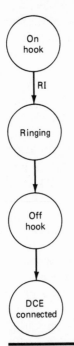

Fig. 3-18 Call-answering connect sequence.

Many microcomputer systems are equipped with manual originate and answer modems with acoustic couplers. With such a DCE operating in answer mode, the RI event causes the user to place the telephone handset into the acoustic coupler to complete the connection between the DCE and the telephone network. When the telephone handset has been taken off-hook, RI no longer cycles between ON and OFF; i.e., the /RI event occurs.

The next important event in the sequence associated with achieving Equipment Readiness is DTR. This event is generated by the DTE to indicate its readiness to exchange data with its local DCE. Notice that DTR must have occurred before the DCE will turn ON the Data Set Ready signal (event DSR). Thus, as with call-origination equipment, the Data Set Ready event implies that the Equipment Readiness phase of operations is complete. The next step is to obtain assurance of the data circuit. This happens when the local DCE senses the transmit carrier from the remote DCE. To indicate that a suitable signal exists for it to commence reception of data off the telephone network, the local DCE turns ON Received Line Signal Detector; i.e., the Data Carrier Detect event takes place. Again, call origination and call answering are similar in that DCD events at both ends of the communications link constitute data-circuit assurance.

It is quite clear from the state-transition diagram in Fig. 3-16 that the two event sequences:

$$RI \rightarrow /RI$$

and

$$/DTR \rightarrow DTR$$

come together at states B_1 and B_2 as prerequisites to generation of the Data Set Ready event by the DCE. A fact that complicates the state-transition diagram (but lends flexibility to the notion of a target-oriented event) is that RI, DTR, and RTS can occur in any one of the six sequences possible for them.

The remainder of the event sequences are quite similar to those for call-origination equipment, so we will not dwell on them here. However, there is one additional point that should be made. Note that for both call-origination and call-answering equipment, it is perfectly legal for the DCE to turn ON Clear to Send before it has asserted its Received Line Signal Detector line. What this means is that CTS may occur *before* the circuit has been assured. The RS-232-C standard very explicitly states that CTS does *not* constitute circuit assurance. The state-transition diagram illustrates this quite nicely.

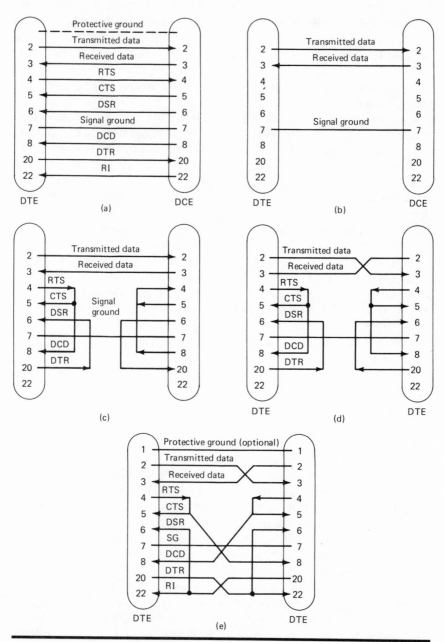

Fig. 3-19 Special cable implementation: (a) full-duplex standard cable,(b) three-wire economy model,(c) three-wire model with luxury loop-back,(d) null modem with luxury loop-back,(e) null modem with double cross.

ADDITIONAL (NONSTANDARD) COMMON CONFIGURATIONS

Figure 3-19 gives schematic diagrams for several RS-232-C special (non-standard) cables that quite often appear in microcomputer systems. Figure 3-19a is the schematic for the full-duplex standard cable that is the subject of the previous few sections. Figures 3-19b to e show variations from the standard. Each diagram depicts two rounded, rectangular connectors labeled DTE and DCE, respectively. The numbers inside each connector correspond to the RS-232-C pin assignments for the circuits shown in the diagram. The circuits are represented by horizontal lines that are labeled with the shorthand signal-event identifiers that we have been using throughout this chapter. (See Table 3-6.) The arrows at the ends of the signal lines indicate the direction of the associated RS-232-C signal. The following paragraphs briefly discuss the rationale behind each of the cables diagramed in Fig. 3-19.

Three-Wire Economy Model

The schematic in Fig. 3-19b is of a cable with only three conductors if the optional Protective Ground circuit on pin 1 is left out. There are many, many microcomputer systems and interfacing configurations for which this cable is entirely adequate.

This cable provides the bare-bones minimum number of circuits necessary for full-duplex communications. Hence, the circuits present are Transmitted Data on pin 2, Received Data on pin 3, and Signal Ground on pin 7. Signal Ground is necessary to provide a common reference for the voltages (i.e., data) present on pins 2 and 3.

The most common pitfall in using a cable such as the three-wire economy model is that many common microsystems components use the Request to Send and Clear to Send circuits. Unfortunately, such equipment will not transmit data until it receives an asserted Clear to Send signal. A large class of equipment that falls in this category uses a USART integrated circuit to implement its serial I/O port. The cable shown in Fig. 3-19c has one set of additions to the economy model that will trick USART-based I/O ports into transmission.

Three-Wire with Luxury Loop-Back

Figure 3-19c shows how the three-wire cable shown in Figure 3-19b can be enhanced with the following loop-back jumpers:

Request to Send . . . jumpered to . . . Clear to Send
Request to Send . . . jumpered to . . . Received Line Signal Detector
Data Terminal Ready . . . jumpered to . . . Data Set Ready

There are more modest implementations (economy loop-back) that involve subsets of the above loop-back jumpers. For example, implementing just the Request to Send–Clear to Send jumper is quite common in microcomputer systems.

The rationale behind the above jumpers can be easily explained in the context of the state-transition logic associated with the RS-232-C standard full-duplex interface. By jumpering Data Terminal Ready to Data Set Ready, the Equipment Readiness phase is completed as soon as the DTE asserts its Data Terminal Ready line. Quite often, this occurs when power is applied to the DTE. When the DTE asserts Request to Send, again an event that is often associated with power-up, the Circuit Assurance phase is immediately completed because Request to Send is jumpered directly to Received Line Signal Detector. In other words, the Request to Send event immediately (and automatically) triggers the Data Carrier Detect event. Since Request to Send is also jumpered to Clear to Send, Request to Send implies immediate completion of the Channel Readiness Phase. So the bottom line is that Data Terminal Ready and Request to Send are the only two events required to achieve the target data-exchange state.

Certainly this cable simplifies the logic associated with obtaining a full-duplex RS-232-C data exchange. What specific features has this simplification omitted? The Data Terminal Ready–Data Set Ready jumper disregards the connect-sequence event. The Request to Send–Data Carrier Detect jumper eliminates the function of the DCE associated with sensing a transmit carrier from a remote modem. The Request to Send–Clear to Send jumper voids any significance associated with Request to Send–Clear to Send handshaking. In particular, the service for preventing overrun errors that is provided by Request to Send–Clear to Send handshaking has been disabled. Notice that all the above features that have been removed by the loop-back jumpers can be associated with the telephone network connection. Therefore, in microsystems possessing no telephone network links, it is reasonable to omit those RS-232-C circuits that are not pertinent.

Null Modem with Luxury Loop-Back and the Null Modem with Double-Cross

The cable schematics shown in Fig. 3-19*d* and *e* both incorporate a crossover of the Received Data and Transmitted Data circuits. As we mentioned earlier, this type of crossover is a very commonplace requirement in microcomputer systems because there are an abundance of DTE microsystem components and a dearth of DCE microsystem components. This means that there is a recurring requirement to trick two DTEs into communicating over a strictly *local* RS-232-C interface (i.e., one with no modems or DCEs). The concept of using a null-modem cable to interpose between two DTEs arose to satisfy this requirement and has been essentially identified with the above-mentioned "crossover technique." There are two standard variants on the basic crossover. One, called the luxury loop-back variant, simply loops back the modem control lines. The other emphasizes the crossover technique more, and has been dubbed the double-cross variant.

The adjective *luxury* refers to the existence of more modest cables in which some loop-backs can be omitted. Quite frequently, none of the loop-backs are present, in which case the cable is just a null-modemization of the three-wire economy model. Notice that the null modem with luxury loop-back (Fig. 3-19*d*) is merely a null-modemization of the three-wire with luxury loop-back model, and utilizes the same system of looping back of the control signals. The only remaining issue is to address the reason for crossing over pins 2 and 3. This is vividly depicted in Fig. 3-5, which illustrates two DTEs attempting to talk on pin 2 and listen on pin 3.

Figure 3-18*e* shows the double-cross variant which involves a crossover between the following two pairs of control signals:

> Request to Send ↔ Received Line Signal Detector
> Data Terminal Ready ↔ Ring Indicator

In addition, this cable employs the following two loop-backs:

> Request to Send . . . jumpered to . . . Clear to Send
> Ring Indicator . . . jumpered to . . . Data Set Ready

Of course, the final important crossover relates to the null-modem function of this cable:

> Transmitted Data ↔ Received Data

Rationale for Loop-Backs and Crossovers

Let us investigate the rationale behind all these loop-backs and crossovers. First, we can dispense with the crossover between Transmitted Data and Received Data. As with any null modem, this crossover resolves the problem of two components listening and/or talking on the same circuits. What is the significance of crossing Data Terminal Ready and Ring Indicator? This crossover has the effect of equating the DTR event with the RI event. This means that when the DTE asserts DTR to indicate that it is ready to exchange data, the answering (normally remote) DTE is immediately given the necessary stimulus (namely, RI) to believe that it has an incoming call. Since Ring Indicator is jumpered directly to the Data Set Ready circuit, the originating DTE is immediately given an asserted Data Set Ready line. The result is that the originating DTE thinks that the Equipment Readiness phase is complete. Conversely, the Equipment Readiness phase at the answering DTE's station will be complete as soon as the answering DTE asserts its Data Terminal Ready line. The next step is for one of the DTEs to assert its Request to Send line. An RTS event from either the originating or answering end immediately makes the opposite end think that the circuit is assured by virtue of the cross-connection between the Request to Send and Received Line Signal Detector circuits. This means that a DTE will not obtain data-circuit assurance until its counterpart generates an RTS event. However, because of the jumper between Request to Send and Clear to Send, RTS generates an immediate and automatic CTS event. This is consistent with the control logic in a standard full-duplex interface, in that Clear to Send has never been intended to imply anything about the state of a counterpart DCE-DTE station. It is interesting to note that the Request to Send–Data Carrier Detect association implements the notion that although a DTE *can* transmit, it perhaps is not *wise* for it to transmit until its correspondent at the other end of the line is prepared to respond. Again, this is consistent with standard full-duplex RS-232-C control logic.

This completes our discussion of common RS-232-C cable configurations. In the next section, we explore the relationship between the RS-232-C standard and the new compatible standards RS-422, RS-423, and RS-449 that have been developed by the EIA.

RS-449: A COMPATIBLE IMPROVEMENT OVER RS-232-C

Background

By 1973, just 4 years after publication of the RS-232-C standard, the EIA realized that the accelerating rate of technological change was going to require major revision to RS-232-C. The drawbacks of the then 4-year-old standard can be summarized as follows:

- The data-transmission rate is limited to 20 kbps.
- The distance for transmission is limited to 50 ft.
- The standard does not adequately specify a connector, resulting in a proliferation of 25-pin designs that are sometimes not compatible with each other.
- Only one conductor per circuit is used, with only one signal return (ground) for both directions of transmission.
- The interface uses *unbalanced* transmitters and receivers. (If you are

unfamiliar with the term *unbalanced,* for now just remember that an unbalanced interface circuit is less desirable (relative to performance) than a balanced circuit. We will discuss these terms in a separate section on the RS-449 interface standard electrical characteristics.
- The interface can generate considerable crosstalk among its component signals.
- The overall interface design is for discrete component technology, not integrated circuit technology.

With these problems in mind, the EIA developed a set of objectives to be achieved in a new standard to replace RS-232-C:

- To maintain compatibility with the old RS-232-C standard: In particular, it should not be necessary to change RS-232-C equipment in order to operate with devices implementing the new standard.
- To support higher data-transfer rates.
- To support longer distances.
- To add interface circuits to perform such functions as loop-back testing.
- To resolve the mechanical interface problems that had resulted from the lack of connector specifications in the RS-232-C standard.
- To improve the electrical characteristics of the interface by providing for balanced circuits.

There were several alternative methods for achieving the above objectives which were considered by the EIA. The first was to create a revision to RS-232-C. This approach was rejected because a revision, RS-232-D, would have to be compatible with RS-232-C to such an extent that several of the other objectives would not be feasible. Thus, in November 1977 the EIA issued a new standard, which directly addresses mechanical and functional characteristics:

> EIA Standard RS-449, "General Purpose 37-Position and 9-Position Interface for Data Terminal Equipment and Data Circuit-Terminating Equipment Employing Serial Binary Interchange."

As part of the effort to improve on the electrical characteristics of RS-232-C, the EIA produced two additional standards:

> EIA Standard RS-422-A, "Electrical Characteristics of Balanced Voltage Digital Interface Circuits," December 1978.
> EIA Standard RS-423-A, "Electrical Characteristics of Unbalanced Voltage Digital Interface Circuits," September 1978.

The EIA has, in essence, "modularized" its approach to developing standards by separating the specifications for electrical characteristics from those for mechanical and functional characteristics.

What will be the impact of these three new standards upon the user community? From the standpoint of performance, the new standards are a distinct improvement, which will encourage manufacturers to begin producing equipment with the enhanced interface. However, because of considerations of cost and convenience, and because of general resistance of the user community to change, it will be some time before the upgraded standards find widespread acceptance. The U.S. government has begun to require RS-449 interfaces on some new equipment, thus initiating the slow but sure transition away from RS-232-C. The manufacturer of the first inexpensive RS-232-C/RS-449 adapter will probably get rich. Expensive adapters exist now. As improved adapter designs are developed, and as they move into mass production, prices will drop and availability will be widespread, and the transition should wind up being rather painless.

Before discussing the EIA standards RS-449, RS-422-A, and RS-423-A, let us investigate what is meant by the terms *balanced* and *unbalanced*.

Balanced and Unbalanced Electrical Circuits

Figures 3-20, 3-21, and 3-22 show schematic diagrams of the circuits associated with RS-232-C, RS-423-A, and RS-422-A, respectively. Underneath the figures are several bulleted items that summarize the most important characteristics of each circuit type. The major advantage of both RS-422-A and RS-423-A electrical characteristics over those of RS-

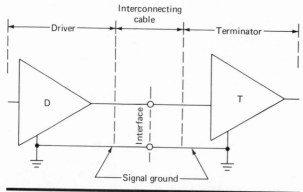

Fig. 3-20   *RS-232-C Electrical Interface Characteristics.* (EIA.)

- *Designed for discrete component technology*
- *Unbalanced interface*
- *Uses one conductor per circuit with one signal return (ground) for both directions*

- *Signal rate limited to ≤20 kbps*
- *Distance limited to ≤15 m*
- *Generates considerable crosstalk*

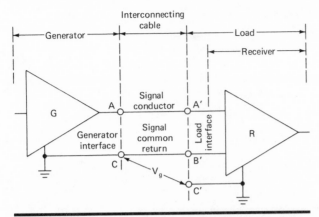

Fig. 3-21 *RS-423-A electrical interface characteristics.* (EIA.)

- *Designed for IC technology*
- *Unbalanced generator*
- *Differential receiver*
- *One conductor per circuit with an independent signal return for each direction*

- *Signaling rate up to 300 kbps. Distance: 1000 m (≤3 kbps) to 10 m (at 300 kbps)*
- *Reduced crosstalk*
- *Interoperable with V.28 and V.11/X.27*

232-C is that they employ *differential inputs.* To understand that this is indeed an advantage, consider the following situation: A DCE and DTE are located in the same building but on different electrical distribution systems. Thus the ground potential for the DTE could, say, be 5 V higher than that for the DCE. Under the RS-232-C standard, in order to send a logic 1, the DTE may assert a voltage of −5 to −15 V, relative to its signal ground. If the DTE asserts any voltage in the range from −5 to −7.9 V relative to its (the DTE's) signal ground, then the DCE will sense a voltage, relative to its (the DCE's) signal ground, in the range from 0 to 2.9 V. The net result is that the DTE may think that it is communicating a logic 1, when, in fact, the DCE is sensing an indeterminate voltage level. The difference in the two signal grounds (V_g in Figs. 3-21, and 3-22) of 5 V is the culprit here.

Another potential hazard in many buildings is electrical noise. The RS-232-C cable may pass through an electromagnetic field which alters the voltage that the DTE placed on the line. If the magnitude of this induced noise is sufficiently large, logic 0s can end up looking like logic 1s, and vice versa.

Problems like the above can be greatly reduced by using differential receivers. The most important characteristic of this type of receiver is that *it measures a difference in voltage between two inputs.* One input is always the conductor that carries the signal of interest. The other input

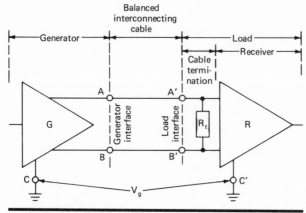

Fig. 3-22 *RS-422-A electrical interface characteristics.* (EIA.)

- **Designed for IC technology**
- **Balanced generator**
- **Differential receiver**
- **Two conductors per circuit**
- **Signaling rate up to 10 Mbps**

- **Distance: 1000 m (≤100 kbps) to 10 m (at 10 Mbps)**
- **Considerably reduced crossstalk**
- **Interoperable with V.10/ X.26**

is the sender's signal ground for RS-423-A. For RS-422-A, the receiver inputs are the conductors that carry the output of a differential signal generator. Note that a differential signal generator transmits a particular voltage level as the difference of two outputs. Referring again to the above problems, note that the first, which stemmed from a 5-V difference in signal grounds, is resolved in the RS-423-A interface by the inclusion of the sender's (DTE's) signal ground as an input to the (DCE's) differential receiver. The RS-422-A circuit solves the problem by not using signal ground at all. What about the noise problem? Since both inputs to the receiver pass through the same electrical environment, they will experience the same voltage alteration, say X volts. Hence the new "noise-laden" voltages at points A' and B' (in Figs. 3-21 and 3-22) are $V_a + X$ and $V_b + X$, assuming that the original voltages were V_a and V_b. Note that the difference measured by the receiver remains the same as the differential voltage originally sent, namely $V_a - V_b$, since $(V_a + X) - (V_b + X)$ is equal to $V_a - V_b$.

For more detailed information on balanced and unbalanced circuits, we refer you to almost any good text on operational amplifiers.

Mechanical Characteristics

Unlike RS-232-C, RS-449 gives detailed specifications for a standard connector. Since the DB-25 connector, so often associated with RS-232-

C interfaces, had performed quite satisfactorily, the EIA decided to use connectors from the same family for RS-449. However, RS-449 incorporates more than 25 signals, so a new connector had to be selected. In order to satisfy the requirements of some foreign standards organizations, the EIA elected to use two connectors: a 37-pin and a 9-pin connector. All signals associated with the basic RS-449 interface requirements appear on the 37-pin connector, while the secondary channel circuits appear on the 9-pin connector.

A second important extension provided by the RS-449 standard for mechanical characteristics addresses the connector size (including protective hood) and the manner in which the connector may be latched to a DTE or DCE. We know of several commercially available microcomputer systems that have panels equipped with precut mounting slots for RS-232-C connectors that are too closely spaced. The result is that only every other slot can be used. If the panel designers had been supplied with standard dimensions for RS-232-C connectors and plastic hoods, this waste never would have occurred. The most common latching arrangement for RS-232-C connectors is via two small screws that go through the plastic connector hood into holes in the DTE or DCE cabinet. A major problem with this design is that a screwdriver is required either to latch or to unlatch a connector. As a consequence, the latching capability is ignored. The weight of the cable plus the connector tends to pull an RS-232-C plug out of its associated socket, with usually undesirable consequences. In contrast, RS-449 mechanical specifications provide for a latch that requires no special tool to engage or disengage. Note that the dimensions of this latch are also covered by the standard to ensure that equipment manufacturers have enough information to allow for them.

Figures 3-23 to 3-26 show the extensive specifications provided for the RS-449 37-pin connector. Table 3-7 gives a list of the signals present in the RS-449 standard with their connector pin assignments. For the subset of signals that are present in an RS-232-C interface, we have included both the 232 signal name and the 449 signal name. The function of some of these signals will be discussed briefly in the section on the functional characteristics specified by RS-449.

Compatibility between RS-449 and RS-232-C is accomplished at the connector level with an adapter that has been specified in an application note published by the EIA:

"Application Notes on Interconnection between Interface Circuits using RS-449 and RS-232-C," *EIA Industrial Electronics Bulletin*, No. 12, November 1977.

The pin assignments for RS-449 interface signals were made to facilitate the adapter design.

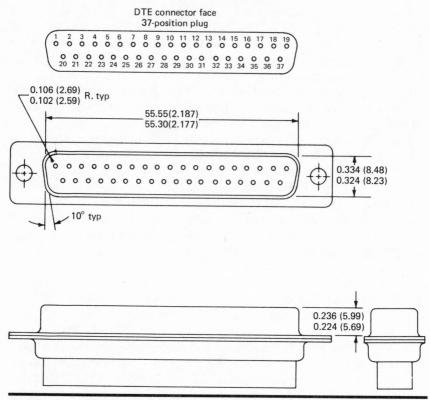

Fig. 3-23   37-position DTE interface connector. Dimensions are given in inches followed by millimeters in parentheses. (EIA.)

The RS-449 standard also contains detailed guidance as to the maximum recommended cable length. Since the range of data transmission rates supported by RS-449 (0 to 2 million bits per second) is so wide, recommended cable length is given as a function of data-transmission rate. The faster the data rate, the shorter the cable should be. Figure 3-27 shows the relationship given by the EIA for "how fast" versus "how far."

Electrical Characteristics

As we mentioned earlier, an important objective of the RS-449 standard has been to maintain compatibility with RS-232-C interfaces. This has been partially accomplished by allowing the use of unbalanced RS-423-A electrical characteristics for the subset of RS-449 circuits that are most commonly used in RS-232-C interfaces, provided the data-trans-

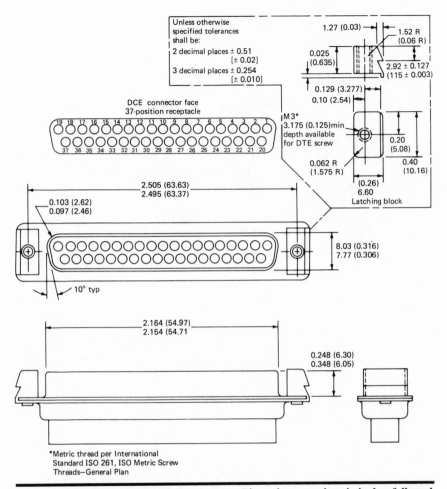

Unless otherwise
specified tolerances
shall be:
2 decimal places ± 0.51
[± 0.02]
3 decimal places ± 0.254
[± 0.010]

1.27 (0.03)
1.52 R
(0.06 R)
0.025
(0.635)
2.92 ± 0.127
(115 ± 0.003)

0.129 (3.277)
0.10 (2.54)

DCE connector face
37-position receptacle

M3*
3.175 (0.125)min
depth available
for DTE screw

0.20
(5.08)

0.062 R
(1.575 R)

0.40
(10.16)

(0.26)
6.60
Latching block

2.505 (63.63)
2.495 (63.37)

0.103 (2.62)
0.097 (2.46)

8.03 (0.316)
7.77 (0.306)

10° typ

2.164 (54.97)
2.154 (54.71)

0.248 (6.30)
0.348 (6.05)

*Metric thread per International
Standard ISO 261, ISO Metric Screw
Threads—General Plan

Fig. 3-24   37-position DCE interface connector. Dimensions are given in inches followed by millimeters in parentheses. (EIA.)

mission rate is less than 20 kbps. Note that 20 kbps is the upper limit for RS-232-C data-transmission rates. This "RS-232-C compatibility" subset of RS-449 circuits, designated as category I, is as follows:

Send Data (SD)

Receive Data (RD)

Terminal Timing (TT)

Send Timing (ST)

Receive Timing (RT)

Request to Send (RS)

Clear to Send (CS)

Receiver Ready (RR)

Terminal Ready (TR)

Data Mode (DM)

All the remaining circuits belong to category II.

The full RS-449 standard stipulates that for data rates under 20 kbps, category I signals may be implemented via unbalanced RS-423-A or balanced RS-422-A electrical characteristics. For data rates over 20 kbps, the category I signals must use the balanced RS-422-A electrical characteristics. The category II signals always use the unbalanced RS-423-A electrical characteristics.

Figures 3-28 and 3-29 show the relationship between voltage and logic state for RS-422-A and RS-423-A, respectively. Note that both standards use a narrower voltage range (-6 to $+6$ V) than RS-232-C (-15 to $+15$ V). In order to compensate for the unbalanced transmitter cir-

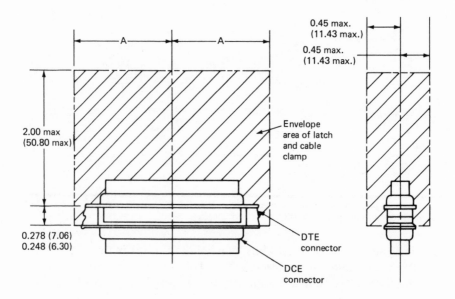

Connector	Dimension A	
	in	mm
37-position	1.500 max.	38.10 max.
9-position	0.742 max.	18.85 max.

Fig. 3-25 DTE connector envelope size. Dimensions are given in inches followed by millimeters in parentheses. (EIA.)

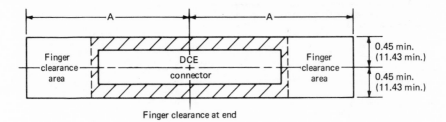

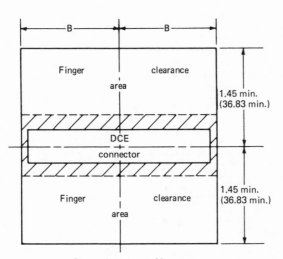

Finger clearance at side

Connector	Dimension A		Dimension B	
	in	mm	in	mm
37-position	2.500 min.	63.50 min.	1.500 min.	38.10 min.
9-position	1.742 min.	44.25 min.	0.742 min.	18.85 min.

Fig. 3-26 DCE connector mounting clearance. Dimensions are given in inches followed by millimeters in parentheses. Shaded area represents maximum area of DTE connector including latch and cable clamp. Finger clearance area may be shared by two connectors. (EIA.)

cuit, the noise margin for RS-423-A is 3.8 V, which is significantly greater than the 1.8-V noise margin for RS-422-A.

Functional Characteristics

Table 3-7 gives a complete listing of the RS-449 circuits and the equivalent RS-232-C circuit where applicable. Notice that the EIA uses new circuit names and two-letter identifiers. The new naming convention

TABLE 3-7 *RS-449 interface circuits and RS-232-C equivalents*

	EIA RS-449		EIA RS-232-C	37-pin connector assignment
SG	Signal Ground	AB	Signal Ground	19
SC	Send Common			37
RC	Receive Common			20
IS	Terminal in Service			28
IC	Incoming Call	CE	Ring Indicator	15
TR	Terminal Ready	CD	Data Terminal Ready	12-30
DM	Data Mode	CC	Data Set Ready	11-29
SD	Send Data	BA	Transmitted Data	4-22
RD	Receive Data	BB	Received Data	6-24
TT	Terminal Timing	DA	Transmitter Signal Element Timing (DTE source)	17-35
ST	Send Timing	DB	Transmitter Signal Element Timing (DCE source)	5-23
RT	Receive Timing	DD	Receiver Signal Element Timing	8-26
RS	Request to Send	CA	Request to Send	7-25
CS	Clear to Send	CB	Clear to Send	9-27
RR	Receiver Ready	CF	Received Line Signal Detector	13-31
SQ	Signal Quality	CG	Signal Quality Detector	33
NS	New Signal			34
SF	Select Frequency			16
SR	Signaling Rate Selector	CH	Data Signal Rate Selector (DTE source)	16
SI	Signaling Rate Indicator	CI	Data Signal Rate Selector (DCE source)	2
SSD	Secondary Send Data	SBA	Secondary Transmitted Data	3
SRD	Secondary Receive Data	SBB	Secondary Received Data	4
SRS	Secondary Request to Send	SCA	Secondary Request to Send	7
SCS	Secondary Clear to Send	SCB	Secondary Clear to Send	8
SRR	Secondary Receiver Ready	SCF	Secondary Received Line Signal Detector	6
LL	Local Loop-Back			10
RL	Remote Loop-Back			14
TM	Test Mode			18
SS	Select Standby			32
SB	Standby Indicator			36

Note that all the category I circuits have two pin assignments to handle differential transmitters and receivers.

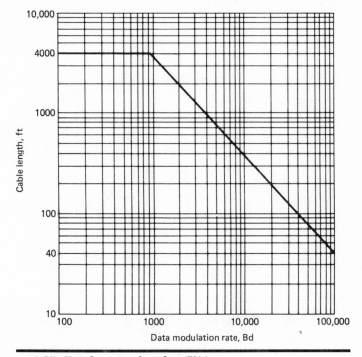

Fig. 3-27 *How far versus how fast.* (EIA.)

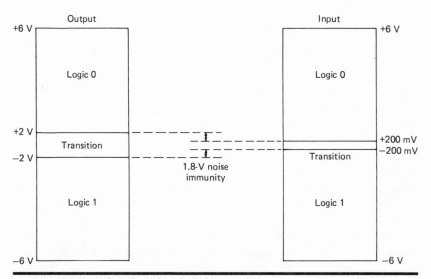

Fig. 3-28 *RS-422-A interface electrical characteristics.*

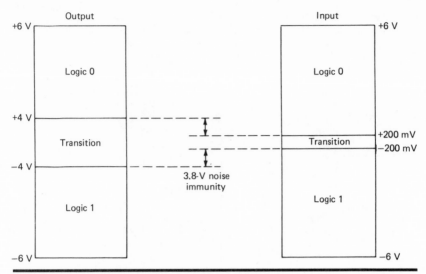

Fig. 3-29 RS-423-A interface electrical characteristics.

was adapted for two reasons. First, the new names more closely reflect the circuit's function, e.g., RS instead of CA for Request to Send. Second, the fact that there are no new two-letter identifiers that match the old RS-232-C identifiers avoids any confusion of the two systems.

There are ten new signals defined in the RS-449 standard (see Table 3-7). Very briefly, their functions are as follows:

- *Send Common (SC):* This circuit provides a signal common return path for unbalanced circuits used in the direction from DTE to DCE.
- *Receive Common (RC):* This circuit provides a signal common return path for unbalanced circuits used in the direction from DCE to DTE.
- *Terminal in Service (IS):* This circuit indicates to the DCE whether or not a DTE is operational. A major application of this circuit is to provide a busy signal on a telephone line associated with a given DTE in a hunt group, if the DTE is out of service.
- *New Signal (NS):* This signal is used mostly in multipoint polling applications. In multipoint polling situations in which the remote DCEs are operating in a switched carrier mode, the control DTE polls each remote DTE, in turn, for a message. The remote DTEs, upon receiving the poll, if they have a message for the control DTE, will immediately respond by placing the message on the channel. The result is that the incoming signals to the control DTE-DCE station appear as a series of short message bursts. Thus the control DCE must accommodate to a rapid succession of discrete messages from the sev-

eral remote stations. In nonsynchronous systems, the Receive Data line may experience some spurious signals as the message source changes. The NS circuit is used by the control DTE to signal the control DCE that the message from one remote DTE has ended and a new one is about to begin. Thus NS enables the control DCE to ignore spurious intermessage signals and still respond properly to actual data signals on the channel.

- *Select Frequency (SF):* This signal is used mostly in multipoint polling applications. SF is used by the DTE to select the transmit and receive frequencies used by its DCE. The selection indicates whether the high or low frequency is for transmit; the unselected frequency is used for reception.
- *Local Loop-back (LL):* This circuit is used by the DTE to request the initiation of local loop-back testing. When LL is activated, data and control signals generated by the local DTE are looped through the local DCE back to the local DTE again. The purpose of this testing is to verify the functionality of the local DTE and DCE.
- *Remote Loop-back (RL):* This circuit is used by the DTE to request the initiation of remote loop-back testing. When RL is activated, data and control signals generated by the local DTE are looped through the local DCE to the remote DCE and back again to the local DTE. The purpose of this testing is to verify the functionality of the local DTE, the local DCE, the communication channel, and the remote DCE.
- *Test Mode (TM):* TM is used to indicate to a DTE when a test condition has been established that involves its local DCE.
- *Select Standby (SS):* This signal is used by the DTE to request a switch to standby equipment to replace the prime equipment. The purpose of SS is to facilitate rapid recovery from equipment failures.
- *Standby Indicator (SB):* SB is used to indicate to the DTE whether regular or standby facilities are in use. Typically, SB is activated in response to activation of the SS circuit.

20-MILLIAMPERE CURRENT-LOOP INTERFACES

The last serial communications physical-layer protocol that we will discuss is the 20-mA current-loop interface. Many computer peripherals and computer serial I/O ports implement 20-mA loops, and several manufacturers treat 20 mA as the baseline interface, handling RS-232-C as an extra cost option. In this section, we briefly discuss the principles behind operation of a 20-mA current-loop communications interface, its advantages and disadvantages, and, finally, how it compares with the other physical-layer protocols that we have discussed: TTL, RS-232-C, RS-422-A, and RS-423-A.

Background and Operation

Until the early 1960s, wire services and military teleprinters used 60-mA loop current to communicate over long distances. The Teletype Corporation Model 32 teletypewriter, introduced in the 1950s, used 20-mA loop current, but it was not until its successor, the Model 33 Teletype, was introduced in 1962 that 20-milliampere interfaces became widely used. For many years, the Model 33 was the dominant commercially available computer terminal. Over half a million Model 33s have been manufactured and sold throughout the world.

Loop-current circuits transfer data in a bit serial manner by turning the current off and on in a time-series sequence to represent logic levels. Loop current can be used for data exchange between any two devices, but usually it is used for mechanical teleprinters. As the name implies, the loop-current interface is configured as a loop in which one or more devices are connected serially. Figure 3-30 illustrates this fact. The current from the source flows out the positive terminal, through each of the devices in the circuit, and back to the source's negative terminal. An interruption at any point in the current loop is felt by all the devices attached to it.

A typical loop circuit for interfacing a computer and terminal consists of two simplex lines, one for each direction of data flow. (See Fig. 3-31.) Theoretically, however, there is no limit to the number of devices that can be connected to a current loop. The limitation on the number of devices is purely a function of the power available to drive the current. Note that each device on the line constitutes a resistance to the

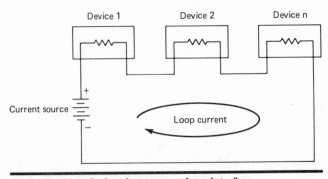

Fig. 3-30 Three devices in a current-loop interface.

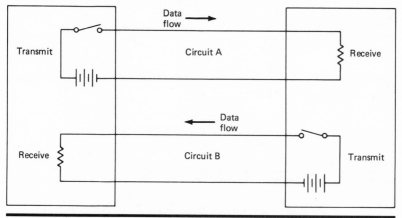

Fig. 3-31   Typical computer terminal 20-mA interface.

current flow. A simple application of Ohm's law shows that the more resistance there is in the loop, the less current will flow. Recall that Ohm's law is the equation

$$E = I \times R$$

where E = voltage, volts
       I = current, amperes
       R = resistance, ohms

Therefore, in order to maintain a constant current I at 20 mA, as resistance increases as a result of additional devices on the loop or extensions in the loop's length, the voltage must be increased to compensate.

Thus, the basic components of a 20-mA current-loop interface are a current source, a current switch (or transmitter), and a current detector (or receiver). As mentioned above, a typical computer-terminal interface consists of two loops: one for the keyboard, in which the computer is a receiver, and one for the display/printer, in which the computer is a transmitter. An important question that must always be considered when connecting two current-loop devices is the location of the current source. If the current source is in the transmitter, then the transmitter is called active. Similarly, an active receiver is a receiver which supplies the current source to the loop.

Typically, the current source is generated by means of a voltage source and a resistor. Again, applying Ohm's law, a voltage source of 20 V will suffice to drive a device with 1000 Ω of internal resistance, since the current I equals 20/1000 or 0.020 A. For a device with 15 kΩ of internal resistance, 0.020 × 15,000 = 300 V must be supplied to maintain 20-mA current in the interface.

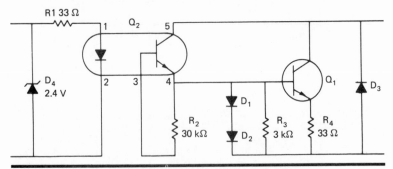

Fig. 3-32 *Optical coupler in a 20-mA current-loop interface.*

As with the DCE-DTE pairing for RS-232-C interfaces, 20-mA current loops require one active and one passive component. Thus, an active transmitter must be interfaced with a passive receiver. Similarly, a passive transmitter must be interfaced with an active receiver.

Also, as with RS-232-C, there is a way around the above pairing requirements, at least for interfacing two active devices. The 20-mA current-loop analog to a null modem is the optical coupler. Figure 3-32 shows how an optical coupler works. An optical coupler consists of a light-emitting diode (LED) and a phototransistor packaged in a light-tight box. The LED acts as a passive receiver. When current from an active transmitter flows through it, the LED emits light, which the phototransistor senses. When the phototransistor senses light, it acts as a passive transmitter in that it starts to conduct the current available to it from the active receiver. The result is that the optical coupler, because it acts as a passive receiver and a passive transmitter, is able to join two active 20-mA devices.

The only way to interface two *passive* devices is to equip one with a power supply.

Advantages and Disadvantages

For applications involving relatively short distances, less than 2000 ft, and data-transmission rates less than 9600 bps, the 20-mA current-loop circuit can usually outperform an RS-232-C interface; 20-mA interfaces are simple and inexpensive. Typically, they are implemented using a DB-25 connector, utilizing pins that are unassigned by the RS-232-C standard. Thus, it is possible to implement both 20-mA current loop and RS-232-C in one cable. The 20-mA interface requires four lines:

Transmit data

Transmit return

Receive data

Receive return

To save the expense of the relatively costly DB-25 connectors, you can use any good-quality four-conductor connector to implement a current-loop interface.

There are two major problems associated with 20-mA current-loop interfaces. The first problem stems from the lack of any mechanical or electrical standard. Consider the following problem. Suppose that you have two devices that are equipped with 20-mA current-loop interfaces. One device is an active transmitter and the other device is a passive receiver. The transmitter device has an internal resistance of 15 kΩ, and has, therefore, been outfitted with a rather hefty power supply to drive it, namely 300 V. (We established the necessity for 300 V above.) The receiver has an internal resistance of only 1000 Ω. If you were to naively plug these two devices together, the result would be a lot of sparks and black smoke. The reason is that the 300-V power supply is too much for the passive receiver. The receiver would be subjected to 300/1000 or 0.3 A of current. This is 15 times what it was designed to operate with! Thus, the lack of standardization in 20-mA interfaces can have devastating consequences.

The second major problem with 20-mA current-loop interfaces is noise. The electric circuits in the teleprinters use rotating contacts or commutators. The mechanical bounce of these contacts results in electrical noise which must be filtered out to ensure reliable communication. Also, over long distances, 20-mA active components tend to introduce crosstalk into other wires that are nearby.

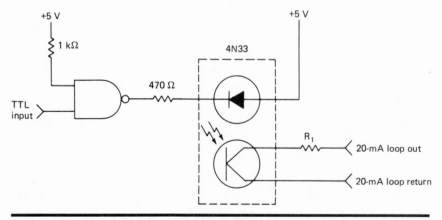

Fig. 3-33 Circuit to convert from TTL levels to 20-mA current.

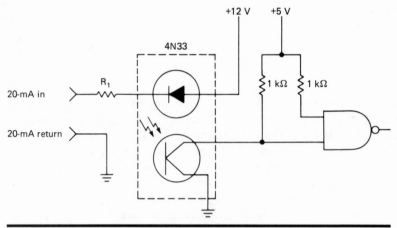

Fig. 3-34   *Circuit to convert 20-mA current to TTL levels.*

Comparison with Other Interface Conventions

For applications requiring the use of loop-current communications, the conversion to and from TTL levels is usually quite simple. Figure 3-33 on the previous page shows a circuit designed to convert from TTL levels to 20-mA loop current. This circuit uses the 4N33 optical isolator to lessen the possibility of overloading sensitive computer circuits. The value of R_1 is determined by the resistance of the line, the connectors, and the internal resistance of the receiving device. The value of R_1 must be such that 20 mA of current is maintained in the circuit. Note that the circuit in Fig. 3-33 assumes that the current source is supplied externally, i.e., that this circuit is for a passive transmitter.

TABLE 3-8   *Performance characteristics of TTL, RS-232-C, RS-422A, RS-423-A, and 20-mA current loop*

	TTL	RS-232-C	RS-422-A	RS-423-A	20 mA
Logic 0 (generator)	0 to 0.4 V	+5 to +15 V	+2 to +6 V	+4 to +6 V	No current
Logic 1 (generator)	2.4 to 5 V	−5 to −15 V	−2 to −6 V	−4 to −6 V	20 mA
Maximum speed, bits/s	~25 × 10⁶	2 × 10⁴	10⁷ 10⁵	10⁵ 900	
Maximum distance, ft	~30	50	40 4000	40 4000	Over 1000

The circuit in Fig. 3-34 shows a 20-mA-to-TTL interface where the interface is the current source. Again, for protection, this circuit uses the 4N33 optical isolator. Just as for the circuit in Fig. 3-33, the value of R_1 is determined by the resistance of the rest of the circuit and must be such that 20 mA is maintained in the circuit.

Table 3-8 summarizes several important performance attributes of TTL, RS-232-C, RS-422-A, RS-423-A, and 20-mA current loop.

EXPERIMENTS

Introduction

The following experiments are designed to accomplish two objectives. The first is to give you experience in fabricating and testing RS-232-C cables. The second is to show you how to build two very useful tools for identification and correction of communications problems. The first tool is a serial-line monitor box that you can use to tap into a communications line and "see" data flowing across it. The line monitor will be used extensively in the experiments at the end of Chapter 4, on asynchronous serial transmission. The second tool is a null-modem box. You will use a null-modem box to configure an RS-232-C cable that is customized to conform to the exact requirements of your microcomputer system.

The experiments on RS-232-C interfacing can be summarized as follows:

Experiment No.	Objective
3-1	To fabricate the three-wire economy model RS-232-C cable.
3-2	To build a line monitor and use it to test microcomputer components. In particular, you will test your terminal and RS-232-C serial I/O port to see if it is a DTE or DCE.
3-3	To fabricate a cable that brings out RS-232-C signals to a 25-pin D-type connector from the edge of a printed circuit card.
3-4	To build a null-modem box and use it to design an RS-232-C cable that is customized to your particular microsystem's needs.

Materials and Equipment

The following equipment is required for the experiments in this chapter:

30-W soldering iron with fine tip

.032 solder

Wire cutters

Ohmmeter

Small saber saw

Screwdriver

Small vise

Computer I/O terminal with keyboard

Several possible sources for the materials used in the experiments are listed below:

Locally from an electronics supply store such as a Radio Shack or from some microcomputer retail stores.

By mail order from:
JADE Computer Products
4901 West Rosecrans Avenue
Hawthorne, CA 90250
(800) 421-5500 for continental United States
(800) 262-1710 for inside California

JAMECO Electronics
1021 Howard Avenue
San Carlos, CA 94070
(415) 592-8097

INMAC
Department MP
49 Walnut Street
PO Box B
Norwood, NJ 07648
(201) 767-3601

INMAC
Department MP
1148 107th Street
Arlington, TX 76011
(214) 641-0024

INMAC
Department MP
130 S. Wolfe Road
PO Box 404
Sunnyvale, CA 94086
(408) 737-7777

Unfortunately, it can be difficult to obtain computer electronics supplies locally. It is worth some effort to locate good local sources, but if none exist, the above sources are reliable and reasonably fast if you order by telephone. In any case, be sure to get on their mailing lists.

Experiment 3-1

Purpose:

The purpose of this experiment is to fabricate the RS-232-C cable called the three-wire economy model.

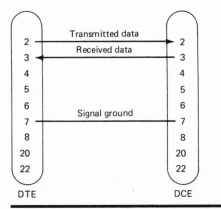

Fig. 3-35 Schematic of three-wire economy model RS-232-C cable.

Schematic Diagram of Circuit

See Fig. 3-35.

Step 1

Obtain from your favorite source the following materials:

Quantity	Item
2	25-pin male D connector, such as CINCH DB-25-P
2	Connector hoods, such as CINCH DB-51226-1
6 ft	4-conductor cable. The length may be longer or shorter, if you wish. The length must not be longer than 50 ft if the cable is to meet the RS-232-C standard.
2 ft	$\frac{1}{16}$-in heat-shrink tubing

Step 2

Plug in the soldering iron to start heating it up.

Cut and strip back about 1 in of the outer jacket on each end of the four conductor cable. Strip back about 1/4 in of the insulation on each of the four exposed conductors at each end of the cable. When you have completed this step, your cable should look like the one shown in Fig. 3-36.

Using the 30-W soldering iron and .032 solder, carefully tin the eight exposed conductor ends. *To tin* means to coat the exposed conductor with a thin layer of solder. The purpose of tinning is to facilitate heat

Fig. 3-36 Stripped four-wire conductor cable.

transfer, thus making it easier to solder the conductors to the 25-pin connectors in subsequent steps.

Carefully tin the solder lugs for pins 1, 2, 3, and 7 on each DB-25 connector. Again, the purpose of this is to facilitate soldering the cable conductors to the connectors.

Cut 8 pieces of heat-shrink tubing into 1/4-in lengths.

Step 3

Your cable is now ready to be attached to the 25-pin connectors. In this step, you will follow the schematic in Fig. 3-35 to implement the Protective Ground circuit on pin 1.

The Protective Ground circuit on pin 1 is optional, but since we have four conductors, we can afford to use it. If one of the four conductors in your cable has green insulation, it should be used for the Protective Ground circuit. In any case, we will refer to the conductor connected to pin 1 as the green one. To connect the Protective Ground circuit on pin 1, use the following procedure:

a. Slide one of the 1/4-in lengths of heat-shrink tubing over the green conductor on one end of the cable.
b. Solder the conductor to pin 1. **CAUTION:** Do not use very much solder. If you are new at this type of work, it may help to have someone else hold the connector and conductor in a convenient position while you apply the solder. There are two major pitfalls to avoid. The first is creating a solder bridge, in which you form an electrical connection between two neighboring pins with an errant blob of solder. The second is creating a cold solder joint, which may appear to be connecting pin to conductor but actually is not. Cold solder joints are caused by moving the conductor or connector while the solder is cooling. In a subsequent step you will test your new cable for both of these faults.
c. Slide the heat-shrink tubing over the connection that you just made. Without touching the tubing, apply heat to it with the soldering iron. The tubing should shrink tightly around the conductor and connector pin within seconds of applying heat. *Do not expose the tubing to heat any longer than necessary.*

Use steps a, b, and c to attach the other end of the green conductor to pin 1 of the second connector. You have now completed connecting circuit AA, Protective Ground.

As we mentioned earlier, Protective Ground is an optional circuit. The reason is that, in some installations (probably not your microsystem installation), ground loops between pieces of equipment are a problem. A ground loop is caused by a difference in electric potential between the ground circuit of two or more pieces of electrical equipment. When these pieces of equipment are connected, a current will flow in the conductor connecting the ground circuits. This current flow can induce noise in the signal circuits, potentially degrading, if not totally disabling, communications. Hence, where ground loops are a problem, pin 1 is left unconnected.

Step 4

In this step you will connect the remaining circuits in the three-wire economy model RS-232-C cable.

First, connect pin 7 using a black insulated conductor, if possible. Pin 7 implements circuit AB, Signal Ground.

Next, connect the circuits on pin 2 (circuit BA, Transmitted Data) and pin 3 (circuit BB, Received Data). This completes connection of all the cable circuits.

Step 5

Having completed an initial implementation phase, let us now move into a quality assurance phase. You will test your cable for three possible faults:

1. Solder bridges
2. Cold solder joints
3. Continuity between connectors

Turn on your ohmmeter and set it so that it displays ohms in the range of 200 and up. Using your ohmmeter, you will test for pin-to-pin continuity, first within one connector, and second between the two connectors. For continuity within one connector, put the ohmmeter probes on the 12 possible pairs of pins from the group 1, 2, 3 and 7. In particular, to test continuity between pins 1 and 2, put the ohmmeter probes on the two pins and look at the reading on the ohmmeter. If the ohmmeter reading says anything less than infinity, the two pins are connected electrically. This means that either there is a bare conductor in contact with the two pins or there is a solder bridge between pins 1 and 2. You must eliminate this because it will disable data transmission on pin 2. You eliminate a solder bridge by removing the solder that is connecting the two pins. Solder suckers and solder wick are two products that can be used to remove the solder.

Test for cold solder joints and continuity between connectors by putting the ohmmeter probes on corresponding pins of the two connectors. For example, put the probes on the two pin 1s and read the ohmmeter. This time a reading of 0 ohms is *good*. It means that the two connectors have an electrical connection at pin 1. Similarly, test the two pin 2s, 3s, and 7s. If any pair does not produce a 0 reading, then the conductor is somehow not making proper contact with the connector. Possible causes are a cold solder joint or just a bad connection. To correct the situation, reconnect the errant pin and conductor.

Step 6

Install the connector hoods on the DB-25 connectors. You now have a quality-assured three-wire economy model RS-232-C cable. Note that the "three-wire" appellation is still operational, since the cable really uses only the wires connecting pins 2, 3, and 7.

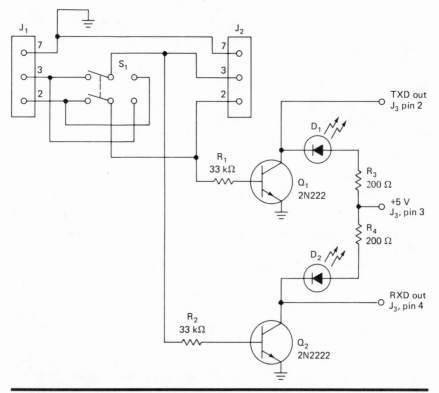

Fig. 3-37 Circuit for hardware line monitor.

Experiment 3-2

Purpose

The purpose of this experiment is to fabricate a hardware interface to be used for monitoring data flow on a serial line of the EIA RS-232-C type. The main function of the hardware monitor is to provide a parallel monitor to tap on a pair of RS-232-C data lines. The data coming from the parallel tap is then converted to TTL logic levels and made available to an output connector. In addition, the monitor provides a pair of indicator LEDs to allow the user to see activity on the serial data lines.

Schematic Diagram of Circuits

See Fig. 3-37.

Step 1

Obtain from your favorite source the following materials:

Quantity	Item
1	Small project box, such as Radio Shack no. 270-231
1	Subminiature DPDT switch, such as Radio Shack no. 275-614
2	25-pin female D-type connectors such as CINCH DB-25-S
2	2N2222 general-purpose transistors
2	33-kΩ ¼-W resistors
2	200-Ω ¼-W resistors
1	0.2-in red LED
1	0.2-in green LED
1	Red LED mounting lens
1	Green LED mounting lens
1	4-conductor connector set (male/female)
1	Spool no. 24 awg solid hookup wire

Step 2

Cut and strip the outer jacket of one end of six 4-in lengths of hookup wire, leaving about 1/4 in of exposed conductor. Solder one end of one 4-in hookup wire to the solder lug for pin 2 of one of the DB-25 female

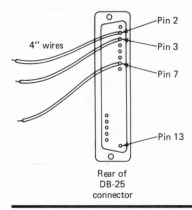

Fig. 3-38 *DB-25 connector with three 4-in hookup wires attached.*

connectors. Similarly, solder two 4-in hookup wires to the solder lugs for pins 3 and 7 of the same DB-25 connector. (See Fig. 3-38.)

Repeat the above for the second DB-25 connector. Mount one connector to each side of the project box. (See Fig. 3-39.)

Step 3

Drill three small holes in the top of the project box. These holes will be used to mount the LED indicator lens and DPDT switch. Figure 3-40

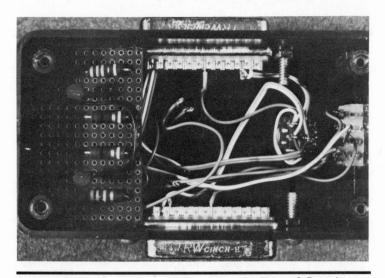

Fig. 3-39 *Hardware monitor assembly—Step 2.* (Dr. Thomas C. Brown.)

Fig. 3-40   Hardware monitor—assembly step 3. (Dr. Thomas C. Brown.)

shows one way to space the holes; however, the placement of these holes is not critical. The size of the holes will depend on the size of the shaft on the switch and the diameter of the LED lens. Next mount the switch and the two lens holders in the appropriate holes.

Step 4

Mount the four-conductor connector in one end of the project box. This connector will provide the incoming power to the monitor and an interface for the TTL outputs.

Step 5

Assemble the circuit shown in Fig. 3-37. The transistors and resistors should be mounted on a small piece of epoxy glass circuit board. Phenolic circuit board may be used, but it is much harder to work with. Use the no. 24 awg hookup wire to make the connections to the switches, the connectors, and the LEDs. The connector being used for the TTL output and the power input should be polarized to eliminate the possibility of improperly connecting power to the circuit. Figure 3-41 shows how we laid out the circuit on a piece of 1.5- by 1.5-in epoxy glass circuit board.

For our monitor, we used the green LED for diode D_1 and the red LED for diode D_2. Pin 1 of the four-pin connector is used for ground, and, hence, the points in the schematic that show a connection to ground are all brought out to pin 1 of J_3 (the four-conductor connector). The pin assignments for the other pins on J_3 are:

Pin 1: Ground

Pin 2: TXD out

Pin 3: $+5$ V

Pin 4: RXD out

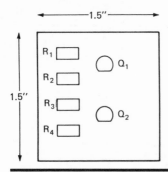

Fig. 3-41   Layout of circuit for hardware line monitor.

Before continuing with the assembly of the hardware monitor, let us investigate the function of the circuit that forms its basis. The entire circuit of the monitor is shown in Fig. 3-37. The circuit diagram can be divided into two major functions. The first is the null-modem circuit that is centered around S_1. The second is the signal-level conversion circuit associated with Q_1 and Q_2.

The null-modem functions allow the user to cross the transmit and receive data lines or to send the data directly through without changing connector pin assignments. To see this let us assume that switch S_1 is positioned such that the center set of poles is making contact with the left set of poles. Now let us trace the path of incoming data from pin 3 of connector J_1 to connector J_2. Our objective is to see whether the data that come into the box on pin 3 of J_1 leave the box on J_2 pin 2 or pin 3 of J_2. Data coming in on pin 3 of J_1 go first to the upper left pole of switch S_1. Assuming that this pole is connected to the upper center pole, the data then proceed along the connection between the upper center pole and pin 3 of J_2. Similarly, we can trace the connection from pin 2 of J_1 to the lower left pole, thence to the lower center pole, and finally out to pin 2 of J_2. Therefore, when the center poles are connected to the left poles, the monitor box acts as a straight pass-through.

What about when the switch S_1 is in the other position? Then the center poles are connected to the right poles. In this case, if we follow the data from pin 3 of J_1 to the right, down, to the right and up to the lower right pole, and then out the lower center pole, we see that they are connected to pin 2 of J_2. Thus pin 3 of J_1 is connected to pin 2 of J_2. Now follow the connection from pin 2 of J_1 over to the right, down, across, and then up to the upper right pole. The position of the switch connects the upper right pole to the upper center pole and then over to pin 3 of J_2. Thus, pin 2 of J_1 is connected to pin 3 of J_2. In effect, the monitor box acts as a null modem when the switch S_1 connects the center poles to the right poles by crossing the data lines. This fully describes the null-modem function of the circuit.

The monitoring and signal-level conversion function is centered at transistors Q_1 and Q_2. Again referring to Fig. 3-37, you can see that pin 2 of J_2 is connected to the base of Q_1 through a 33-kΩ resistor. This resistor drops the RS-232-C +12-V and −12-V levels to approximately +1 V and −1 V. This voltage swing then acts to turn Q_1 off and on. An incoming +1 V will turn Q_1 on (or, in other words, cause it to conduct). An incoming −1 V will turn Q_1 off (or cause Q_1 to stop conducting). This switching action will cause the ground connected to the emitter of Q_1 to be felt at the collector when Q_1 is on and a high resistance to be felt at the collector when Q_1 is off. The output labeled *TXD out* is connected directly to the collector of Q_1 and will be at a signal level of 0 V when Q_1 is on. Recall that +12 V on the input of pin 2 of J_2 causes Q_1 to conduct, and that RS-232-C logic level 0 is represented by a signal level of +12 V. Thus, the output at TXD is at TTL logic 0 when pin 2 of J_2 is at RS-232-C logic 0. When TXD is at logic 0, then diode D_1 is forward-biased and will glow, indicating a logic 0 on the data line.

On the other hand, when pin 2 of J_2 is at RS-232-C logic 1, or has a potential of −12 V, then Q_1 is turned off. In this case, the TXD output will be at a signal level of approximately 3.75 V. This signal level comes from the +5 V at the junction of R_3 and R_4. The combined voltage drop of R_3 and D_1 is about 1.25 V, leaving a potential of 3.75 V at the emitter of Q_1. This 3.75 V is a TTL logic level 1 at the TXD output. Also, in this case, D_1 is reverse-biased (or not adequately forward-biased) and will not glow, indicating a mark or logic 1 condition on the RS-232-C data line.

Transistor Q_2 reacts in the same manner to changing levels on pin 3 of J_2. Diode D_2 and output RXD bear essentially the same relationship as D_1 and TXD and monitor the activity on pin 3 of J_2. If two RS-232-C-type devices are connected to J_1 and J_2, the diodes D_1 and D_2 will indicate the flow of data between the devices and the outputs TXD and RXD will provide a TTL output of the data.

Step 6

Complete the assembly of the hardware monitor by mounting the circuit board that you just assembled in the project box, and label the connectors J_1 and J_2. Also label the two switch S_1 positions as S for straight through and X for crossed.

Step 7

The next step is to test your new diagnostic instrument by using it to identify DTEs and DCEs. For this, you will need, in addition to the hardware monitor, the three-wire RS-232-C cable that you fabricated in Experiment 1 and a computer terminal with an RS-232-C interface.

To supply power to your monitor box, connect the +5-V and ground leads from the four-conductor connector to a 6-V battery, or, if one is available, a 5-V power supply.

Next, use the three-wire RS-232-C cable to connect the terminal to connector J_2 on the monitor box. Power up the terminal and observe the LEDs on the monitor box as you type a few keys on the keyboard. You should see one of the LEDs blink. Which LED blinks tells you whether you have a DCE or a DTE. Before you can figure out which one your terminal is, you must know the position of switch S_1. To see the effect of switch S_1, flip it back and forth as you are typing keys at the terminal keyboard. You should see that the blinking LED alternates with each change in the switch setting.

To determine if your terminal is a DCE or DTE, place the switch S_1 in the S position. If the terminal is a DTE, which LED should blink? The answer is that D_1 (the *green* LED) should blink. Conversely, if the terminal is a DCE, D_2 (the *red* LED) should blink.

Step 8

Testing a serial I/O port on your microcomputer is easy if you can execute a program to output a long series of bits to it. The basic strategy is to connect the port to the monitor box with the three-wire RS-232-C cable while it is sending serial data. If the I/O port that you wish to test is the console port which you must use to initiate the program, simply initiate the program and quickly remove the cable from the back of the terminal and plug it in to the monitor box.

Note that the really clever way to determine whether a working console port is a DCE or DTE is to check the terminal, rather than the port. If the terminal is a DTE and the cable that it works with is straight through (i.e., connects pin 2 to pin 2 and pin 3 to pin 3), then the console port *must* be a DCE.

Experiment 3-3

Purpose

The purpose of this experiment is to build three very useful RS-232-C cables. The first cable is used to bring the RS-232-C signals out from the edge of a printed circuit card to a 25-pin female D-type connector. Although this cable will cost you no more than about $20 to $25 to fabricate, and take you no more time than about 30 minutes, it is not

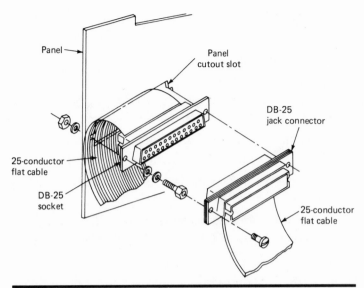

Fig. 3-42   Ribbon cable implemention of an RS-232-C interface.

unusual to see cables of this type sold commercially for as much as $130! The other two cables that you will fabricate in this experiment are for use with the null-modem box that is the subject of Experiment 3-4.

Schematic Diagram of Circuit
See Fig. 3-42.

Step 1
Obtain the following materials:

Quantity	Item
4	Male DB-25 insertion displacement connectors with strain relief and cover, such as JADE CND-5252
1	Female DB-25 insertion displacement connector with strain relief and cover, such as JADE CND-5251
15 ft	26-conductor ribbon cable, such as JADE WCR-261N010
1	Connector that is appropriate for mating with your printed circuit card serial I/O connector.

Note You most probably will need one of the two following 26-pin connectors:

- Card edge, such as JADE CNE-5102620
- Socket, such as JADE CNF-62200

Alternatively, you can order a 26-conductor ribbon cable with the card edge or socket connector already attached:

- 26-pin card-edge connector attached to 36 in of 26-conductor ribbon cable, such as JADE APP-924063-36

- 26-pin socket connector attached to 36 in of 26-conductor ribbon cable, such as JADE APP-924003-36

If you obtain the semiprefabricated cables, subtract 3 ft from the 26-conductor ribbon cable requirement of 15 ft, and purchase only 12 ft.

Step 2

First you will make the cable that brings out the RS-232-C card-edge signals to the female 25-pin connector. You must connect the socket or card-edge connector to the ribbon cable. This is accomplished by using a vise to "mash" the cable to the bank of pins on the connector that must come in contact with the conductors in the ribbon cable. (See Fig. 3-43.)

There are two pitfalls in making the connection between a connector of the insulation displacement type and a ribbon cable. First, make sure that the ribbon cable is properly aligned before you begin to tighten the vise. Second, do not apply too much pressure as you mash together the cable and connector. As soon as the part (labeled U contact in Fig. 3-43) has clicked into place, you have squeezed enough. (Note that if you pur-

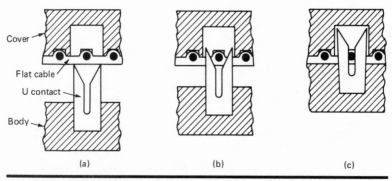

Cover
Flat cable
U contact
Body

(a) (b) (c)

Fig. 3-43 Insulation displacement connection to ribbon cable: (a) positioned, (b) piercing insulation, (c) connected.

chased a semiprefabricated cable, then the above step has already been completed for you.)

Next, connect the female 25-pin connector to the 26-conductor ribbon cable. Note that you must be careful here to match up the two pin 1s. Note also that one conductor in the ribbon cable will not be connected to a pin on the 25-pin D-type connector. By referring to the documentation supplied with your RS-232-C serial I/O card, you should be able to discern which ribbon cable conductor is to be left out. *Read your serial I/O port interface documentation very carefully before you attach the female 25-pin connector* to ensure that you have it aligned properly. Then mash it into place with the vise.

Step 3

To make the first male-to-male ribbon cable, use 5 ft of ribbon cable and two male insertion displacement connectors. After you have attached the first connector, be sure to attach the second so that their pin 1s align and so that the same conductor is left unconnected at both ends. Repeat the process for the second male-to-male cable.

Step 4

You can test the cables that you have just built for continuity and shorts by using an ohmmeter and the technique that is described in Experiment 3-1.

Experiment 3-4

Purpose

The purpose of this experiment is to build another very useful tool for diagnosing RS-232-C communications problems: a *null-modem box*. The tool is called a null-modem box because it implements the null-modem function of crossing and jumpering together various combinations of circuits within an RS-232-C interface. Of the tools used for incorporating null-modem functions into an RS-232-C interface, the null-modem box is the most flexible. This experiment illustrates this flexibility by showing you how to use your newly constructed null-modem box to design an RS-232-C cable customized to meet your system's unique needs.

Schematic Diagram of Circuit

See Fig. 3-44.

Step 1

Obtain the following materials:

Quantity	Item
2	25-pin female D-type connectors, such as CINCH DB-25-S
1	Electronics project box, approximately 5 × 3 × 2 in
1	22- to 26-pin experimenter's socket such as Radio Shack no. 276-175
1	Spool no. 24 awg solid conductor hookup wire

Step 2

Install the two DB-25-S connectors on opposite sides of the project box, with their solder lugs facing inside the box. Mount the experimenter's socket on the bottom of the box. The null-modem box should now resemble the one shown in Fig. 3-45.

Step 3

Cut twenty-five 3-in lengths of the no. 24 awg hookup wire. Strip 1/8 in of insulation from one end of each of the wires.

Starting on the bottom row, pin 25 (please see "note" below) of one of the DB-25-S connectors, solder a 3-in wire to its solder lug. Repeat this process for each of the 11 remaining bottom row pins. Next, cut, strip, and dress the wire on each DB-25 pin so that it can be inserted into the nearest end row of tie points on the experimenter's socket. *Dressing* refers to the process of making every effort to keep the wires from turning into a jungle. Trim the wire to the exact length required, and then arrange the 12 wires neatly as they span the space between solder lug and experimenter's board tie point. Figure 3-46 shows a good example of how the wires should be dressed.

Note   If the experimenter's socket that you purchased has less than 25 tie points, then do *not* tie the RS-232-C unassigned pins to it. These are pins 9, 10, 11, 18, and 25. Instead, just leave them unconnected, or connect them straight through to the counterpart pin on the other DB-25 connector.

Step 4

Next you must connect the top row of DB-25 connector pins to the experimenter's socket. Solder a 3-in length of hookup wire to each of the 13 pins in the top row, i.e., pins 1 through 13. Then, starting with pin 13, cut, strip, and dress the wires and insert them into the remaining tie points in the outer row on the same side as in step 3.

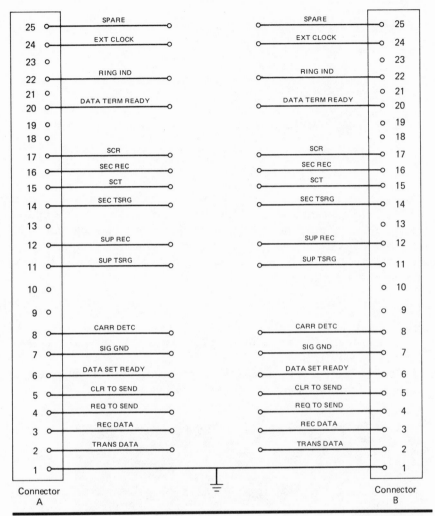

Fig. 3-44   Circuit for null-modem box.

Step 5

Now you have completed the connections to the experimenter's socket for one connector. You are almost half finished. Before you begin to repeat steps 3 and 4 for the second connector, read the following caution.

Caution   You must make sure that the wire coming from pin *n* of one connector is inserted into a tie point in the socket diametrically opposite

Fig. 3-45 *Null-modem box assembled.* (Dr. Thomas C. Brown.)

to the tie point that the other connector's pin of the same number is tied to. *This is not straightforward!* Since the connectors are on opposite sides of the box, pin 1 on one connector is directly opposite pin 13 on the other. Therefore, you will have to dress the wires connecting the second connector to the socket *differently* than the way you dressed those of the first connector.

Step 6

Cut twelve 1- to 2-in lengths of the no. 24 awg hookup wire. Strip 1/4 in of the insulation from each end. (You are getting proficient at wire stripping, yes?)

Referring to Fig. 3-47, make the appropriate jumpers between tie points on each side of the experimenter's socket.

Fig. 3-46 *Null-modem box assembled showing dressed wires.* (Dr. Thomas C. Brown.)

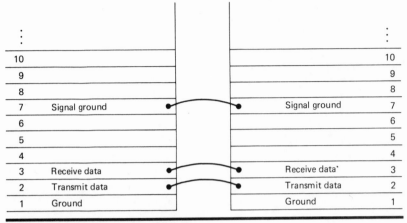

Fig. 3-47 Three-wire economy model jumpers.

Step 7

You are now ready to use the null-modem box to design an RS-232-C cable that is tailored to your system's unique requirements. Using the ribbon cable that you fabricated in Experiment 3-3, bring out the RS-232-C signals at the edge of your console serial I/O port printed circuit card to a female 25-pin D-type connector. Use one of the male-to-male ribbon cables to connect the female null-modem box connector to the female connector attached to the console port. Use the other male-to-male ribbon cable to connect your console terminal to the other side of the null-modem box. You have now brought all the RS-232-C signals from the console port and console terminal to opposite sides of the null-modem box. (See Fig. 3-48.)

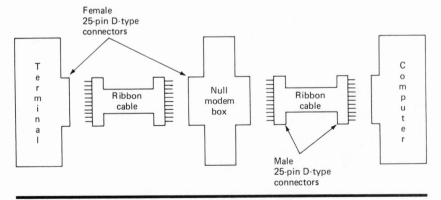

Fig. 3-48 Using a null modem to design an RS-232-C cable.

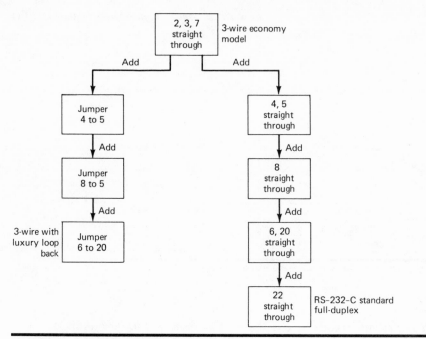

Fig. 3-49 Design tree for straight RS-232-C cable.

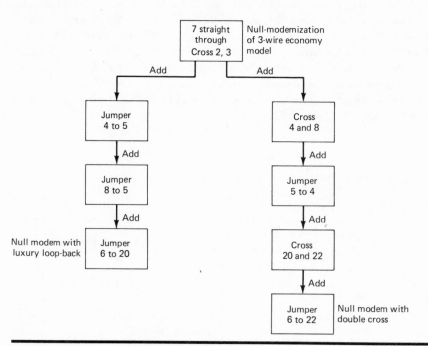

Fig. 3-50 Design tree for null-modem cable.

Step 8

The null-modem box is jumpered to implement the three-wire economy-model cable. To see if such a modest cable is sufficient for your needs, turn on the computer and the terminal and attempt to initiate communications between them. If they can exchange data, then you are done. If nothing happens, then try crossing pins 2 and 3 by appropriately rejumpering the null-modem box. If there is still no communication, your cable design will have to be a bit more sophisticated. In this chapter you have read about several common RS-232-C cable implementations that are useful in microcomputer systems. The two design trees that appear in Figs. 3-49 and 3-50 describe one way to systematically test candidate conductor combinations to design a workable cable. The end result may be exactly the same as one of the cables discussed earlier, or it may be a slight variation of one.

The first step in using the design trees is to determine which one (Fig. 3-49 or 3-50) is appropriate. This is where the hardware monitor box is useful. You can use it to determine which of the following four possible situations describes yours:

I/O port	*Terminal*
DTE	DTE
DTE	DCE
DCE	DTE
DCE	DCE

A DTE/DTE or DCE/DCE combination dictates that you use Fig. 3-50 to design your null-modem cable. A DTE/DCE or DCE/DTE combination means that you should use Fig. 3-49.

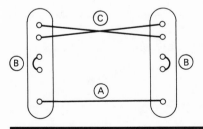

Fig. 3-51 *Cable design options:* (A) *straight through,* (B) *jumper,* (C) *cross.*

Step 9

Use the appropriate design tree to configure a workable RS-232-C cable. Note that both trees offer you a choice between one of two end results (or some subset implementation thereof). There is no special way of making that choice. For your own information, we recommend that you try both. It may well happen that both approaches produce a working cable.

Note also that there are basically three types of connections that you can make in the course of designing a cable:

STRAIGHT THROUGH: Just connect pin n on one connector to pin n on the opposite connector.

JUMPER: Jumper two pins together, e.g., jumper together pins 4 and 5 on each side of the socket.

CROSS: To cross pin x with pin y connect pin x of connector 1 to pin y of connector 2, and connect pin y of connector 1 to pin x of connector 2.

These options are illustrated in Fig. 3-51.

Via further experimentation, you may be able to reorder and/or refine the design steps shown in Figs. 3-49 and 3-50 to produce a more streamlined design approach. In our opinion, the following parable says it all about connector problems:

> There once was a kingdom that was threatened by a wicked, fire-breathing dragon. The king sent out his mightiest knight in white shining armour to conquer the foe. Unfortunately, a faulty nail caused the knight's horse to lose a horseshoe. The loss of a horseshoe caused the horse to go lame. The lame horse caused the knight to lose the critical battle. And, finally the loss of the critical battle caused the king to lose his kingdom. (McNamara, 1977)

What can we say? It all came down to a connecter problem.

Reference

McNamara, John, E., *Technical Aspects of Data Communications*, Digital Press, 1977.

4 *Asynchronous Serial-Related Conventions and Experiments*

The intent of this chapter is to provide a discussion of topics that tie together the material presented earlier in the book and to describe a rudimentary but highly useful hardware-software tool, which we call the *line monitor*. The line monitor allows a user to see, in real time, what is happening on a low-speed communications channel, within a resolution of 1 bit time.

In order to recognize the characters observed on a communications line, one must be aware of the character encoding technique used. Thus, the first section of this chapter presents a discussion of the American Standard Code for Information Interchange (ASCII). This section covers not only the standard ASCII conversion tables, but also provides some interesting information about the function and significance of the ASCII control codes and the critical design issues that must be resolved when developing something as fundamental as a standard for character encoding.

The next major topic addressed is the concept of speed matching or flow control. The purpose of our discussion is to describe the most commonly used speed-matching or flow-control techniques. The connection between speed matching and the ASCII character set is that the most common asynchronous serial speed-matching techniques use ASCII control characters.

The final section is devoted to documenting the hardware and software used to implement the line monitor. Experiment 4-1 at the end of the chapter contains step by step instructions for using the line monitor to observe asynchronous serial protocol as well as for ASCII character encoding.

Experiments 4-2 through 4-5 address the basic problem of Input/Output (I/O) timing control. The techniques investigated range from no timing controls at all to interrupt-driven I/O.

OBJECTIVES

When you have completed this chapter, you will be able to do the following:

- Read and use several conversion tables that show hexadecimal and decimal representations for the ASCII character set.
- Identify ASCII graphic and control characters and state the function of the most widely used control characters.
- Describe several important design issues that were considered by the American National Standards Institute (ANSI) in defining the ASCII standard.
- Describe the XON/XOFF, ETX/ACK, and Reverse Channel techniques for speed matching over serial interfaces.
- Describe the Ready and Strobe signals used to accomplish speed matching for parallel data transfers.
- Use a simple line monitor hardware-software-based tool for observing data transfers on a low-speed serial communications channel.
- Use techniques such as wait loops, polling, and interrupt-driven I/O to implement serial interface device drivers.

AMERICAN STANDARD CODE FOR INFORMATION INTERCHANGE (ASCII)

An important aspect of microcomputer communications that many users often take for granted is the character encoding scheme that underlies all interpretation of the transmitted data. The most universal character set encountered by microcomputer system users is the ASCII character set. ASCII is an acronym for American Standard Code for Information Interchange. ASCII is a 7-bit coded character set that is supported by the American National Standards Institute (ANSI).

The latest version of the ASCII character set is formally referred to as the ANSI X3.4-1977 standard. This is a revision of the earlier ANSI X3.4-1968 standard. The ASCII code set is continually under review, so that rational modifications can be made to it as changing requirements and technology necessitate such modifications. Clearly, any changes to such a fundamental standard as a code set have to be closely controlled to maintain a stable basis for information exchange.

Almost all American manufacturers of computers and computer-related equipment support the ASCII code set, and many of these vendors support the ASCII code set exclusively. Even foreign manufacturers provide equipment that subscribes to the ASCII character set. This widespread acceptance and use of the ASCII character set has had a very positive effect upon the growth of computer communications and the standardization of computer equipment. Can you imagine what it would be like if manufacturers all used their own unique character sets? Users would be forced to purchase every component of a computer configuration from the same vendor, or, alternatively, software would have to be written and incorporated into nonhomogeneous systems to perform code conversion. This situation currently exists in many IBM configurations in which ASCII terminals are connected to IBM EBCDIC systems. In addition to allowing users to purchase and easily interface components from multiple vendors, industry standardization on the ASCII code set has removed a potentially significant barrier to the creation of computer networks. The implementation of communications subsystems to support data transfers between heterogeneous sets of computer equipment has been greatly facilitated by the fact that most systems at least recognize the same characters.

The following paragraphs summarize the ANSI X3.4-1977 standard and appendixes. We gratefully acknowledge the ANSI's permission to reprint parts of this document in the discussion that follows.

The ASCII Character Set

The ASCII character set is a group of 128 distinct 7-bit patterns that each represent some one unique character. It is natural to ask why ANSI chose a 7-bit code. In the appendixes of the ASCII standards document, the statement is made that seven is the minimum number of bits that can support most applications involving information interchange. It is our conjecture that ANSI determined that a reasonable definition of a minimal character set should include 26 uppercase letters, 26 lowercase letters, 10 numerals, and some special characters such as the comma, period, colon, semicolon, etc. Already the total is over 64 characters, which is the number of different characters that 6 bits can support. If the lowercase alphabetic characters are sacrificed, however, a

6-bit code will suffice. Since 7 bits can handle 128 different characters, a 7-bit character set not only accommodates both upper- and lowercase alphabetic characters, but also supports several control codes. We will discuss the ASCII control codes in great detail later in this chapter. Thus, by defining a 7-bit code, ANSI decided to support lowercase characters and control codes.

The above reasoning might justify the selection of a 7-bit code over a 6-bit code, but why not an 8-bit code? The 7-bit code provides for all of the common symbols. In addition the ASCII character set provides for code extensions. If 7 bits provide adequate support, there is no reason to add another bit.

The 128 characters of the ASCII set are shown in Table 4-1.

To determine the ASCII character being represented by a particular bit pattern, the pattern must first be broken down into two parts. The first of these consists of the three most significant bits, the high-order bits 4, 5, and 6. The second part consists of the low-order or least significant bits, 0, 1, 2, and 3. The two parts are used to determine the column and row of the matrix of Table 4-1. The ASCII character that is represented by the bit pattern appears at the intersection of the designated row and column. For example, if we assume that the most significant bit is the leftmost bit, the bit pattern 1000001 breaks into the two parts 100 and 0001. This is depicted in Figure 4-1.

TABLE 4-1   ASCII character matrix

Least significant bits (3, 2, 1, 0)	Most significant bits (6, 5, 4)							
	000	001	010	011	100	101	110	111
0000	NUL	DLE	SP	0	@	P	`	p
0001	SOH	DC1	!	1	A	Q	a	q
0010	STX	DC2	"	2	B	R	b	r
0011	ETX	DC3	#	3	C	S	c	s
0100	EOT	DC4	$	4	D	T	d	t
0101	ENQ	NAK	%	5	E	U	e	u
0110	ACK	SYN	&	6	F	V	f	v
0111	BEL	ETB	'	7	G	W	g	w
1000	BS	CAN	(	8	H	X	h	x
1001	HT	EM	)	9	I	Y	i	y
1010	LF	SUB	*	:	J	Z	j	z
1011	VT	ESC	+	;	K	[	k	{
1100	FF	FS	,	<	L	\	l		
1101	CR	GS	-	=	M	]	m	}
1110	SO	RS	.	>	N	^	n	
1111	SI	US	/	?	O	—	o	DEL

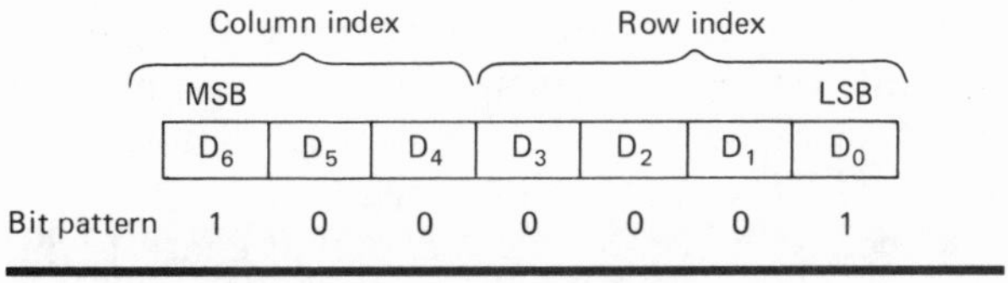

Column index			Row index			
MSB						LSB
D_6	D_5	D_4	D_3	D_2	D_1	D_0

Bit pattern   1   0   0   0   0   0   1

Fig. 4-1   Breakdown of the ASCII bit pattern.

Each of the 8 columns of Table 4-1 are associated with 1 of the 8 possible combinations of the 3 high-order bits of an ASCII character. The 16 rows of the table correspond to the 16 possible combinations of the 4 low-order bits of a character. Since the high-order part of 1000001 is 100, the fifth column holds the character in question. Since 0001 is the low-order part of 1000001, the second row of the table must hold the character. There is only one character that is both in column 5 and in row 2, namely A. Thus, the 7-bit sequence for the letter A is 1000001 or hex 41. Similarly, you can verify that the character m has the ASCII code 1101101.

Categories of ASCII Codes

In defining the ASCII character set, ANSI has divided the codes into two major categories: *graphic characters* and *control characters*. In this context, the term *graphic* means a printable character and not a symbol used for creating computer pictures. The first two columns of Table 4-1 and the DEL character constitute the control character set. The remaining characters constitute the graphic character set. The ANSI standard defines a *control character* as a character whose occurrence in a particular context initiates, modifies, or stops an action that affects the recording, processing, transmission, or interpretation of data. A *graphic character* is defined in the standard as a character, other than a control character, that has a visual representation, normally handwritten, printed, or displayed. The following are examples of the graphic characters:

ABCD. . .XYZ   123. . .90   1!''#$%&'( )* = ` ~?>< +}{] [

ANSI defines three categories of ASCII control characters in the X3.4-1977 standard: *communication controls*, *format effectors*, and *information separators*. The standard defines a *communication control character* as an ASCII character intended to control or facilitate transmission of information over communication networks. An example of a communication control character is the End of Transmission character EOT (hex 04). EOT is used to indicate that the transmission is complete and that no further characters will follow for the current transmission. This character is appended to the data being transferred as the last character. When the destination device receives this character, it knows that no additional information will follow on the channel.

A *format effector* is defined as a control character that controls the layout or positioning of information in printing or display devices. An example of a format effector is the Line Feed character LF (hex 0A). LF will cause the active printing position to be advanced to the same character position on the next line. If you transfer a text message composed of the characters a, b, c, d, LF, e, f, g, then the receiving device will print the message as:

abcd
  efg

An *information separator* is defined as a control character that is used to separate and qualify information in a logical sense. The information separators allow the transfer of variable-length records over a communications channel. For example, if you desire to transfer five records of different lengths, you can transfer an additional record to define the record format or you can, by prior agreement, separate the records with the Record Separator character RS (hex 1E).

The above three control character categories are not all-inclusive. Some control characters do not logically belong to any of the three cat-

---

TABLE 4-2   *ASCII control characters and their categories*

	Control characters	Category*		Control characters	Category*
NUL	Null	CC	DC1	Device Control 1	—
SOH	Start of Heading	CC	DC2	Device Control 2	—
STX	Start of Text	CC	DC3	Device Control 3	—
ETX	End of Text	CC	DC4	Device Control 4	—
EOT	End of Transmission	CC	NAK	Negative Acknowledge	CC
ENQ	Enquiry	CC	SYN	Synchronous Idle	CC
ACK	Acknowledge	CC		End of Transmission	
BEL	Bell	—	ETB	Block	CC
BS	Backspace	FE	CAN	Cancel	—
HT	Horizontal Tabulation	FE	EM	End of Medium	—
LF	Line Feed	FE	SUB	Substitute	—
VT	Vertical Tabulation	FE	ESC	Escape	—
FF	Form Feed	FE	FS	File Separator	IS
CR	Carriage Return	FE	GS	Group Separator	IS
SO	Shift Out	—	RS	Record Separator	IS
SI	Shift In	—	US	Unit Separator	IS
DLE	Data Link Escape	CC	DEL	Delete	—

*CC, communication control; FE, format effector; IS, information separator.

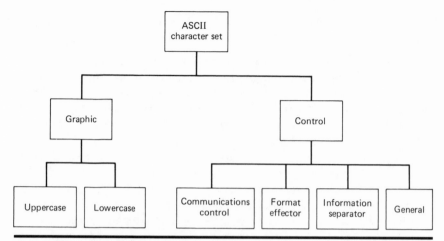

Fig. 4-2   Categories of ASCII codes.

egories. To obtain an all-inclusive characterization, we have created a catch-all fourth category denoted by dashes in Table 4-2. The control characters, their nomenclature, and the subcategory into which each falls are shown in Table 4-2.

The graphic character category is not divided into subcategories by ANSI. However, there is a natural subdivision into two subcategories that is worth noting: *alphanumeric uppercase* and *alphanumeric lowercase*. The uppercase alphanumeric subcategory consists of the graphic characters in the third, fifth, and sixth columns of Table 4-1. The lowercase subcategory consists of the characters in the fourth, seventh, and eighth columns of Table 4-1. Figure 4-2 summarizes the structure of ASCII categories and subcategories, as we have defined them, in a hierarchy chart.

It is common practice to refer to the characters in the ASCII graphic category that are not numeric or alphabetic as special characters. This classification does not, however, result in a clear breakdown in terms of the bit pattern representation of these characters.

Meaning of the ASCII Characters

ANSI has assigned a specific interpretation to each of the ASCII characters. The control character semantics are more interesting and less universally known, so we will discuss them first.

Control Characters

Each ASCII control character addresses a particular control task. The following paragraphs address the control characters, one by one.

NUL: Null   A control character used to accomplish media fill or time fill.

NUL characters may be inserted into a stream of data without affecting the information content of the stream. However, the addition or removal of these characters may affect the information layout or the control of the equipment. One common use of the NUL character is as a pad character to follow carriage returns and line feeds on mechanical teleprinters. Since mechanical printers require some time to move the carriage from the right to the left side of the page after a carriage return, NUL characters can follow the CR to provide this idle time, since the NUL character is nonprintable and requires no response on the part of the receiving device. Depending on the speed of the printer mechanism, one, two, or as many as three of these pad characters may be required. Printers and devices that have an incoming data buffer and use a line protocol to control the incoming data do not need these pad characters.

The relationship between the Break signal and the ASCII character NUL should be noted. The ASCII character NUL consists of seven logic 0s in 1 data byte while the Break signal is defined to occur if a signal line is held in the logic 0 (spacing) state for some time period. This time period usually ranges from 200 ms to a couple of seconds. A Break signal could thus look like the following:

00000000000000000000000000000000000000000000000000

However, since there are no stop bits, the character framing has been lost. Consequently, it is not sufficient to characterize a Break signal as simply the reception of a long sequence of NUL characters. It must be characterized as the reception of a long sequence of NUL characters *together* with the presence of a framing error.

SOH: Start of Heading   A communication control character used as the first character of a message header block.

STX: Start of Text   A communication control character that precedes a text and is used to terminate a heading.

ETX: End of Text   A communication control character that terminates a text.

EOT: End of Transmission   A communication control character used to indicate the conclusion of a transmission, which may have contained one or more texts and associated headings.

To see how these first few communication control characters are used, let us look at an example. Suppose that you wish to send a message to a terminal that has a network address of 12. The text of the message is "Happy New Year." The character count of the message is 14 bytes. By using a simple protocol, the transfer to the terminal from the host device might look something like the following:

SOH,12,STX,H,A,P,P,Y, ,N,E,W, ,Y,E,A,R,ETX,14,EOT

In the above the message the SOH tells the receivers on the network that a message is about to be transmitted. The 12 is the address of the intended receiver. Receivers that have different addresses will know that the message is not intended for them so they can ignore the rest of the message. The terminal that has the address 12 will decode the rest of the message. The STX indicates the end of the header, which only contains the receiver address, and the beginning of the text itself. The text is then transmitted and followed by an ETX character. The ETX character is followed by a character count that provides some degree of error detection by telling the receiver how many characters it should have received. The message is then terminated with the EOT character. The EOT tells the receiver that the transmission is complete and notifies the other devices on the network to begin monitoring the line again for a new address and message.

While the protocol used in this example is primitive, it shows how the message can be divided into different segments by the control characters to allow the receiver to interpret the transmission better. In the more sophisticated protocols used in networks today, the header and the message trailing sections will contain much more control information about the message being transferred. The header might contain a destination address, a source address, a message segment number, and a message number. The trailing section might contain one or more cyclic redundancy check (CRC) characters for error detection purposes.

Let us continue with our investigation of the ASCII control characters.

ETB: End of Transmission Block A communication control character used to indicate the end of a block of data for communications purposes. ETB is used for blocking data when the block structure is not necessarily related to the processing function.

ENQ: Enquiry A communication control character used in data communication systems as a request for a response from a remote station. It may be used as a "Who are you?" (WRU) to obtain identification, or it may be used to obtain status, or both.

ACK: Acknowledge A communication control character transmitted by a receiver as an affirmative response to a sender.

NAK: Negative Acknowledge A communication control character transmitted by a receiver as a negative response to the sender.

ENQ, ACK, and NAK are usually housekeeping characters for the communications protocol being used. They can be invoked by the lower levels of the protocol and are normally transparent to the user.

BEL: Bell A control character for use when there is a need to call for attention. It may control alarm or attention devices, usually audible in nature.

BS: Backspace A format effector that moves the active print position backward in the same line.

HT: Horizontal Tabulation A format effector that advances the active position to the next predetermined character position in the same line.

If we send the message a, TAB, b, TAB, c, TAB, d and the receiving device has a predetermined tab setting at every eighth character position in a line, then the receiving device would print the message in the following manner:

a   b   c   d

Without the TAB characters the message would be printed

abcd

LF: Line Feed A format effector that advances the active position to the same character position in the next line.

When appropriate, this character may have the meaning "New Line." New Line advances the active position to the first character position on the next line. Thus New Line is equivalent to the sequences LF CR and CR LF, where LF has its usual meaning of advancing the active position to the same character position on the next line and CR is the Carriage Return character that is discussed below. Use of the New Line convention requires agreement between the sender and the recipient of data. An example of the use of the LF character was discussed at the beginning of this section.

VT: Vertical Tabulation A format effector that advances the active position to the same character position on the next predetermined line. When agreed upon between interchange parties, VT may advance the active position to the first character position on the next predetermined line.

FF: Form Feed A format effector that advances the active position to the same character position in a predetermined line of the next form or page. When agreed upon between interchange parties, FF may advance the active position to the first character position on a predetermined line of the next form or page.

CR: Carrage Return   A format effector that moves the active position to the first character position of the same line.

SO: Shift OUT   A control character that is used in conjunction with Shift IN to extend the graphic character set.

Currently available serial dot-matrix printers sometimes uses the SO character to shift into enlarged or enhanced character sets. Some systems use the SO to invoke an alternate set of special graphic symbols or characters. These special characters might be graphic representations of chemical, electrical, or civil engineering symbols, for example.

SI: Shift IN   A control character that is used in conjunction with Shift OUT to extend the graphic set. It may reinstate the standard meanings of the bit combinations which follow it.

DLE: Data Link Escape   A communication control character that will change the meaning of a limited number of contiguously following characters. It is used exclusively to provide supplementary data-transmission control functions.

DC$_1$, DC$_2$, DC$_3$, DC$_4$:   Device Controls   Control characters for the control of ancillary devices associated with data processing or telecommunications systems—in particular, with switching devices on or off.

A common use of the DC$_1$ and DC$_3$ characters is to control the incoming data to serial character printers that have a limited buffering capability. That is, some printers can accept data at a greater rate than they can dispose of that data. This imbalance can be maintained only for a limited time period since the buffer is finite. Consequently some method of stopping the incoming data is required while the mechanical printing process catches up. Printers that recognize and use DC$_1$ and DC$_3$ in this manner normally have a one- to five-line buffer for the incoming data. When this buffer becomes partly full, the printer will transmit a DC$_3$ character, usually referred to as XOFF, or Transmit OFF, to notify the host to stop transmitting data until the print mechanism has printed some number of the characters in the buffer. When the printer has reduced the character count in the buffer to some specific level, the printer issues a DC$_1$, usually referred to as XON, or Transmit ON, to the host device to notify the host to begin transmitting data again. A more detailed discussion of this buffer control technique can be found in the second section of this chapter.

SYN: Synchronous Idle A communication control character used by a synchronous transmission system in the absence of any other character to provide a signal from which synchronism may be achieved or retained.

CAN: Cancel A control character used to indicate that the data with which it is sent are in error or are to be disregarded. The specific meaning of this character must be defined for each application.

EM: End of Medium A control character that may be used to identify the physical end of a medium, the end of the used portion of the medium, or the end of the wanted portion of data recorded on a medium. The position of this character does not necessarily correspond to the physical end of the medium.

SUB: Substitute A control character that may be substituted for a character that is determined to be invalid or in error.

ESC: Escape A control character intended to provide supplementary characters (code extension). The escape character itself is a prefix affecting the interpretation of a limited number of contiguous bit patterns.

FS, GS, RS, US: File Separator, Group Separator, Record Separator, and Unit Separator Separators that may be used with data in optional fashion, except that FS is the most inclusive, then GS, and then RS and US is the least inclusive.

DEL: Delete A character used primarily to erase or obliterate an erroneous or unwanted character in punched tape.

DEL characters may also be used as media fillers or time fillers since DEL is a nonprintable character and does not require any response from the device. DEL characters may be inserted into or removed from a stream of data without affecting the information content of that stream. However, the addition or removal of these characters may affect the information layout or the control of equipment, or both.

The ASCII control characters provide essential support for many character-oriented and bit-oriented link-level protocols. In this sense, it is quite reasonable to characterize the ASCII code set as a communications code.

Graphic Characters

The graphic characters are not assigned any special meaning by the ANSI standard. The interpretation of these codes is left to the user.

The space character is one graphic that is sometimes assigned some control significance. Other than representing a blank character position in a display or printout, the space character is often interpreted as a delimiter in much the same fashion as the format effectors and/or information separators.

Some Useful ASCII Conversion Tables

In using the ASCII code set on certain microcomputer systems, certain system-specific conventions render the ASCII to binary code conversions given in Table 4-1 almost useless. For example, the BASIC interpreter used with the Radio Shack TRS-80 model I, level I and II interpreters requires that a programmer specify decimal integer numbers when invoking string functions for the ASCII code set. Thus, a table that converts between ASCII characters and their associated decimal integers is quite useful. Moreover, since most microcomputer assemblers permit the use of hexadecimal notation for specifying byte constants and strings, a hexadecimal to ASCII conversion table can often be very useful. Table 4-3 shows the decimal, binary, and hexadecimal (assuming that the eighth data bit is always low) representations for each of the 128 different ASCII characters. In those cases in which there are no printable graphics for the control characters, the standard control character mnemonic is given.

TABLE 4-3 *ASCII conversion table*

Decimal	Binary	Hexa-decimal	ASCII	Decimal	Binary	Hexa-decimal	ASCII
0	0000000	00	NUL	18	0010010	12	DC2
1	0000001	01	SOH	19	0010011	13	DC3
2	0000010	02	STX	20	0010100	14	DC4
3	0000011	03	ETX	21	0010101	15	NAK
4	0000100	04	EOT	22	0010110	16	SYN
5	0000101	05	ENQ	23	0010111	17	ETB
6	0000110	06	ACK	24	0011000	18	CAN
7	0000111	07	BEL	25	0011001	19	EM
8	0001000	08	BS	26	0011010	1A	SUB
9	0001001	09	HT	27	0011011	1B	ESC
10	0001010	0A	LF	28	0011100	1C	FS
11	0001011	0B	VT	29	0011101	1D	GS
12	0001100	0C	FF	30	0011110	1E	RS
13	0001101	0D	CR	31	0011111	1F	US
14	0001110	0E	SO	32	0100000	20	SP
15	0001111	0F	SI	33	0100001	21	!
16	0010000	10	DLE	34	0100010	22	"
17	0010001	11	DC1	35	0100011	23	#

TABLE 4-3 *ASCII conversion table (Continued)*

Decimal	Binary	Hexa-decimal	ASCII9	Decimal	Binary	Hexa-decimal	ASCII9
36	0100100	24	$	82	1010010	52	R
37	0100101	25	%	83	1010011	53	S
38	0100110	26	&	84	1010100	54	T
39	0100111	27	'	85	1010101	55	U
40	0101000	28	(	86	1010110	56	V
41	0101001	29	)	87	1010111	57	W
42	0101010	2A	*	88	1011000	58	X
43	0101011	2B	+	89	1011001	59	Y
44	0101100	2C	,	90	1011010	5A	Z
45	0101101	2D	-	91	1011011	5B	[
46	0101110	2E	.	92	1011100	5C	\
47	0101111	2F	/	93	1011101	5D	]
48	0110000	30	0	94	1011110	5E	
49	0110001	31	1	95	1011111	5F	
50	0110010	32	2	96	1100000	60	←
51	0110011	33	3	97	1100001	61	a
52	0110100	34	4	98	1100010	62	b
53	0110101	35	5	99	1100011	63	c
54	0110110	36	6	100	1100100	64	d
55	0110111	37	7	101	1100101	65	e
56	0111000	38	8	102	1100110	66	f
57	0111001	39	9	103	1100111	67	g
58	0111010	3A	:	104	1101000	68	h
59	0111011	3B	;	105	1101001	69	i
60	0111100	3C	<	106	1101010	6A	j
61	0111101	3D	=	107	1101011	6B	k
62	0111110	3E	>	108	1101100	6C	l
63	0111111	3F	?	109	1101101	6D	m
64	1000000	40	@	110	1101110	6E	n
65	1000001	41	A	111	1101111	6F	o
66	1000010	42	B	112	1110000	70	p
67	1000011	43	C	113	1110001	71	q
68	1000100	44	D	114	1110010	72	r
69	1000101	45	E	115	1110011	73	s
70	1000110	46	F	116	1110100	74	t
71	1000111	47	G	117	1110101	75	u
72	1001000	48	H	118	1110110	76	v
73	1001001	49	I	119	1110111	77	w
74	1001010	4A	J	120	1111000	78	x
75	1001011	4B	K	121	1111001	79	y
76	1001100	4C	L	122	1111010	7A	z
77	1001101	4D	M	123	1111011	7B	{
78	1001110	4E	N	124	1111100	7C		
79	1001111	4F	O	125	1111101	7D	}
80	1010000	50	P	126	1111110	7E	
81	1010001	51	Q	127	1111111	7F	DEL

TABLE 4-4   *Key sequence for ASCII control characters*

Control character	Control key	Alternate key	Control character	Control key	Alternate key
NUL	CTL @		DLE	CTL P	
SOH	CTL A		DC1	CTL Q	
STX	CTL B		DC2	CTL R	
ETX	CTL C		DC3	CTL S	
EOT	CTL D		DC4	CTL T	
ENQ	CTL E		NAK	CTL U	
ACK	CTL F		SYN	CTL V	
BEL	CTL G		ETB	CTL W	
BS	CTL H	Backspace	CAN	CTL X	
HT	CTL I	Tab	EM	CTL Y	
LF	CTL J	LF	SUB	CTL Z	
VT	CTL K		ESC	CTL [	Escape
FF	CTL L		FS	CTL /	
CR	CTL M	Return, CR, enter	GS	CTL]	
SO	CTL N		RS	CTL >	
SI	CTL O		US	CTL -	

In addition to Table 4-3 another very handy table for those who have ASCII terminals is Table 4-4, which shows how to generate the control characters with the control shift key. The first column of Table 4-4 contains the control character mnemonic, while the second column gives the control key sequence used to generate that character. The third column shows any alternative key (if one exists) that can be used to generate the control character. Note that the alternative methods for generating these control characters are widely accepted conventions, but are not part of the ANSI standard.

As is evident from Table 4-4, there are several control functions that can be generated by more than one key sequence on the keyboard. This is done for the convenience of the operator. Since it takes two keys to generate a standard control character, some of the more frequently used characters are implemented on a single key. Examples of this are the carriage return and escape keys. Some of the more expensive terminals designed specifically for use in communication applications provide a unique key for the complete ASCII character set, including the control characters. Some of these communication terminals even provide a set of graphic symbols for the control characters.

Code Set Design Issues

In designing something as basic as a binary character encoding scheme, there are many critical design issues that must be closely analyzed, since their ultimate resolution will have far-reaching consequences. Some of the factors that must enter into the decision process are the need for an adequate number of graphic and control characters, the decision about whether all of the bit combinations should be defined (thus producing an unambiguous code set), error control issues, compatibility with existing conventions and code sets, structure and categories of codes, impact upon programming language source code representation, keyboard conventions, and collating conventions. The ANSI committee that designed and revised the X3.4-1977 standard considered each of these issues, and we have mentioned some of them earlier in this discussion. Two of particular interest are the keyboard conventions and collating sequence issues.

Keyboard Design

When designing the ASCII code set, the ANSI committee made every attempt to create a code that would be easy to implement on a terminal keyboard. Note that in Table 4-1 the only difference between the characters in the fifth and the seventh columns is the logic level of bit 5. Similarly, the codes in the sixth and the eighth columns differ only in the logic level of bit 5. This is no coincidence. The codes in the fifth column are the shifted (or uppercase) versions of those in the seventh column, and the letters in the sixth column are the shifted (or uppercase) versions of those in the eighth column. Moreover, if you have an ASCII keyboard available, note that the third and the fourth columns have the same relationship—the uppercase symbol associated with each character in the fourth column appears in the same row in the third column.

Collating Sequence

Because of its significance with respect to sorting and searching functions, the collating sequence was an especially critical consideration in the design of the ASCII code set. The ordering or sequencing of individual characters relative to one another, referred to as the collating sequence of the character set, is determined by the value of their binary patterns. Thus the NUL character in ASCII has been assigned the first sort or collating position, while the DEL character will always appear last. The Space occupies the first alphabetic character position in the collating sequence to ensure that the string Musson K. R. precedes Mussonson K. R. Note that the other graphic separators, such as the comma and hyphen, have been encoded to achieve a similar result. Thus, the string Nichols, J. C. appears before the string Nicholson, J. C. in the ASCII collating sequence.

It is interesting to note that the ANSI committee has allowed some flexibility in the standard with regard to collating sequence, being aware that some applications may have unique sequencing requirements. Thus, provided that both communicating parties agree, the collating sequence of the ASCII characters can be changed, while still remaining compliant with the standard.

Related Standards and References

As you can see the ASCII character set is a lot more involved than it first may seem. For the interested reader who would like to pursue this subject further, we suggest the following related standards:

> ANSI X3.41-1974, "Code Extension Techniques for Use with the 7-Bit Coded Character Set of the American Standard Code for Information Interchange"
>
> ANSI X3.15-1976, "Bit Sequencing of the American Standard Code for Information Interchange in Serial-by-Bit Data Transmission"
>
> ANSI X3.16-1976, "Character Structure and Character Parity Sense for Serial-by-Bit Data Communication in the American Standard Code for Information Interchange"
>
> ANSI X3.28-1976, "Procedures for the Use of the Communication Control Characters of the American Standard Code for Information Interchange in Specified Data Communication Links"
>
> ANSI X3.57-1977, "Structure for Formulating Message Headings for Information Interchange Using the American Standard Code for Information Interchange for Data Communication System Control"

An application of some of the ASCII control codes that is quite useful in many microcomputer configurations is speed matching. In particular, for serial 55 character per second (cps) printers, two speed-matching conventions known as XON/XOFF and ETX/ACK use ASCII control codes to match the 1200 bps data rate on the printer's receive data line with the slower 55 cps speed of its print mechanism. The next section provides detailed information on the XON/OFF and ETX/ACK conventions.

SPEED-MATCHING (FLOW-CONTROL) CONVENTIONS

A speed-matching or flow-control mechanism is required by devices that can support a communication data rate that is higher than the rate at which the device can sink the data (dispose of the data) itself. If a bathtub is filled at a rate of 10 gal/min (communication data rate) but is emptied at the rate of 5 gal/min (device sink rate), then some form of speed-matching or flow-control is required. Under these conditions of constant input and output rates, the time at which the tub will over-flow simply depends on the size of the tub (buffering capacity).

The purpose of this section is to discuss techniques that exist for providing a speed-matching or flow-control capability for both parallel and serial interfaces.

A typical example of the necessity for a flow-control mechanism is a printer with a 1200 bps communications link capability and a 55 character per second maximum printing capability. At 55 characters per second on an asynchronous serial line, the printer can sink only about 600 bps yet the instantaneous input rate is approximately double this sink rate at 1200 bps. A typical printer has only a limited buffering capability (256 bytes, for example). Consequently, some provision must be made to turn the input data flow on and off, depending upon the state of the device's print buffer.

Speed Matching for Parallel Data Transfers

Speed matching in parallel data transfers is much simpler than in serial data transfers. A parallel interface typically adds two control signals that implement "handshaking" between the source and the destination communicating devices. First, we will discuss the handshaking protocol that characterizes speed-matching in parallel transfers, and then we will provide a brief description of the Z80-PIO support for this handshaking.

Protocol

The handshaking that takes place between a source and a destination in a parallel data transfer can be characterized by the following scenario:

DESTINATION: I am requesting the next byte, please.
SOURCE: OK, here it is.
DESTINATION: Next byte, please.
SOURCE: Here it is.
•
•
•
(Silence ensues while destination gets ready to read next byte.)
•
•
•
(At last, the silence is broken.)
DESTINATION: Next byte, please.
SOURCE: Here it is.

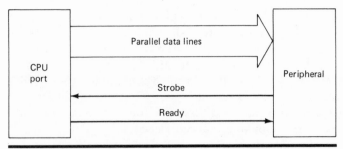

Fig. 4-3   Speed-matching signals for a parallel interface.

The most important characteristic to notice about this handshaking protocol is that it is driven by the destination. In many cases the destination is an I/O device which cannot process the data as fast as the source can supply it. The source cannot transfer a byte until the destination asks for one. Thus, explicit handshaking signals are exchanged between the two correspondents on a character-by-character basis.

Z80-PIO Speed-Matching Support

Almost all parallel I/O ports use the same mechanism to implement speed matching—two control lines in addition to the data lines. These two control lines can be called by different names, but they almost always perform the same function. In keeping with our Z80 microcomputer system model, we will use the signal names that Zilog (as well as many other vendors) has assigned, namely *Strobe* and *Ready*.

Figure 4-3 shows a parallel interface that implements a Strobe-Ready line. This interface could be implemented by the Z80-PIO circuit or by a similar circuit.

The Strobe signal is controlled by the destination and therefore is used to request the next data byte. The Ready signal is controlled by the source and therefore serves to notify the destination that the next byte is ready. For more detailed information on such topics as signal timing, Z80-PIO programming, and specific software functions that must be performed, we refer you to the following:

Zilog Z80-PIO Technical Manual
Z80 Microprocessor Interfacing, book 2, by J. C. Nichols, E. A. Nichols, and P. R. Rony, Howard W. Sams & Co., Inc., Indianapolis, 1978.

Speed Matching for Serial Data Transfers

The implementation of the function performed by the Strobe and Ready lines for parallel data transfers is somewhat more complicated for serial data transfers. For asynchronous serial data transfers the bit rate or baud rate is established in advance of a data transfer and is fixed throughout the complete data transfer. The transmitter clocks bits out on the serial link at this bit rate, and the receiver *must* be able to receive 10 or 11 bits in each data byte at this bit rate. There is relatively little room for speed differences here. If a serial speed-matching protocol is implemented, the receiver does not necessarily have to maintain such a possibly torrid pace for a very long time. The following scenario characterizes the type of handshaking that must occur.

DESTINATION: I'm Ready.
SOURCE: Here is data A.
SOURCE: Here is data B.
SOURCE: Here is data C.
SOURCE: Here is data D.
DESTINATION: I'm not Ready.
•
•
•
(Silence ensues while source awaits Ready message from destination.)
•
•
•
DESTINATION: I'm Ready.
SOURCE: Here is data E.
SOURCE: Here is data F.

The above scenario is similar to parallel handshaking in that the destination turns the transfer on and off via some type of signal. The most important difference is that the handshaking does not occur on a character-by-character basis in the serial case. The source pumps out characters at the agreed-upon bit rate until it receives a request to stop from the destination.

An interesting consequence of the serial speed-matching technique shown above is a "pipeline effect." When the destination decides that it is getting too far behind and sends out a request to stop, the flow of bits across the interface will not cease for possibly several bit times. First, there are bits on the line that were sent before the destination issued the stop request, and, second, more bits will be placed on the line by the source in the time interval between the destination sending the stop request and the source receiving it. Thus, the destination must anticipate that the stream will continue for a while, even after it issues its Not Ready signal. Similarly, the destination must sense the fact that its free buffer space is getting low so that it can signal the source to resume transmission of the data. The destination does not want to be in the position of waiting for data if the source has data to send.

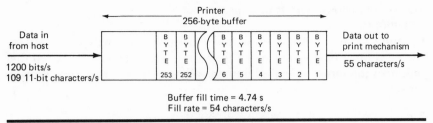

Fig. 4-4   *Letter-quality-printer speed matching with a buffer.*

Example 1

The following specific example illustrates the events that characterize serial speed-matching problems. Many letter-quality character printers are capable of interfacing to 1200 bps transmission links, but can only print at the rate of 45 to 55 characters per second. This makes sense because a device that can print 45 to 55 characters per second must be fed at the rate of 450 bps (10 bits per character at 45 characters per second) to 605 bps (11 bits per character at 55 characters per second). This bit rate falls between two popular serial bit rates, namely, 300 and 1200 bps. To use the letter-quality printer at 300 bps is to restrict a slow device to run at an even slower speed. However, in order to use the 55 character per second printer with a 1200 bps input bit rate, the input data at 1200 bps must be buffered at the printer to allow it to print at its own rate, even though data are coming in at a faster rate. Figure 4-4 illustrates the situation.

At least two crucial questions come to mind: first, how big should the buffer be, and, second, what should the printer do when the buffer is full? A typical buffer size is 256 bytes. A 1200 bps serial input rate corresponds to a character (byte) input rate of 109 bytes per second if 11 bits are transmitted for each 8-bit data byte. A 55 character per second output rate means that the buffer is filling at the rate of 54 characters per second. Thus a buffer size of 256 bytes could buffer data for approximately 4.74 seconds without overflowing. This is not a very long time period.

Example 2

Many printers buffer a line at a time. Due to the print mechanism, a line is not printed until the end of line character is received, at which point the entire line is printed. This is typical of many dot-matrix printers. While the same control mechanisms are used, the buffering itself is based on lines being printed as opposed to being based on the buffer capacity.

Serial Transfer Speed-Matching Techniques

Creating a bigger buffer will not solve the problem; it will only postpone the inevitable, *buffer overflow*. A flow-control mechanism is needed that will turn the input faucet off and on. We will discuss three such mechanisms:

- XON/XOFF control character method
- ETX/ACK control character method
- Reverse Channel signal method

These three speed-matching control techniques are listed in order of most common usage. The XON/XOFF method is the most common, with the Reverse Channel technique being much less common.

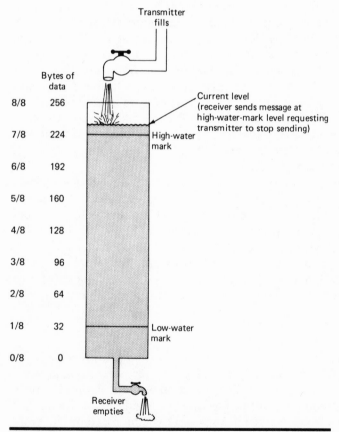

Fig. 4-5 Receiver buffer with 256-byte capacity at high-water mark.

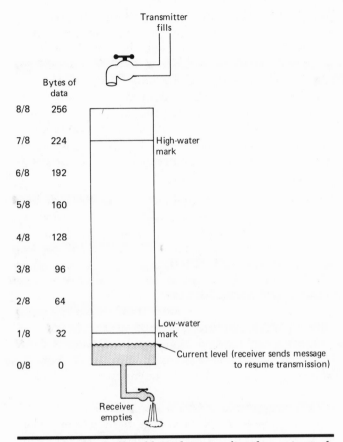

Fig. 4-6 *Receiver buffer with 256-byte capacity at low-water mark.*

XON/XOFF   The XON/XOFF method is diagrammed in Figures 4-5 and 4-6. There are three "water marks" associated with the XON/XOFF method. The high-water mark is typically set at the seven-eighths-full level of the buffer, while the low-water mark is typically set at the one-eighth-full level. The current water mark indicates the current level of data in the buffer. During data transfer, the transmitter fills the buffer, while the receiver, working simultaneously, empties the buffer. Since the buffer filling rate is greater than the buffer emptying rate, the buffer begins to fill up. When the current water mark reaches the high-water mark, the receiver sends the XOFF ASCII control character to the transmitter. The XOFF control character is Device Control 3 (DC$_3$) ASCII control character, with hexadecimal value 13. The transmitter must implement appropriate software to recognize the DC$_3$ and cease transmitting. Once the transmitter stops, the buffer begins emptying at the rate of 55 characters per second. When the current water mark crosses the low-water mark, the receiver sends to the transmitter an XON ASCII control character, Device Control 1 (DC_1), whose hexadecimal value is 11. The transmitter software must recognize the DC_1 and resume transmission. Thus XON/XOFF requires driver software in the transmitter to recognize these two control characters and take action based on the receipt of either one of them. In addition the method requires some intelligence in the receiver to recognize when the current water mark crosses the high-water mark or when it drops below the low-water mark.

ETX/ACK   The ETX/ACK speed-matching technique is less common than the XON/XOFF method in newer devices. ETX/ACK handshaking requires both the transmitter and receiver to be able to recognize the receipt of a special ASCII control character. The transmitter must be able to recognize the ASCII control character ACK (Acknowledge), hex 06, and the receiver must have the capability to recognize the ASCII control character ETX (End of Text), hex 03.

Under the ETX/ACK control technique, the transmitter sends a predetermined fixed-size block to the receiver, followed by the ETX character. The receiver empties its buffer and, upon recognizing the ETX control character, sends to the transmitter the ASCII control character ACK, whereupon the transmitter sends a subsequent fixed-size block to the receiver.

The XON/XOFF method is slightly more efficient since the receiver is never waiting for data, while with the ETX/ACK method the receiver incurs a slight delay while waiting for data after the ACK has been sent and before the first characters of the next block have arrived from the transmitter. This efficiency difference is insignificant for many applications.

Reverse Channel   The Reverse Channel method is similar to the XON/XOFF method except that the control messages (Stop and Resume) are conveyed back to the transmitter through the use of a separate signal line rather than being handled as ASCII characters sent from the destination to the source.

A Reverse Channel signal is typically implemented on the Secondary Request to Send line of the RS-232-C interface cable. This signal appears on pin 19 of the usual DB-25 RS-232-C connector. The signal direction is from the device that we have referred to above as the receiver (the printer in Examples 1 and 2) to the device referred to above as the transmitter (the computer in Examples 1 and 2).

In the language of the RS-232-C standard the Reverse Channel is a

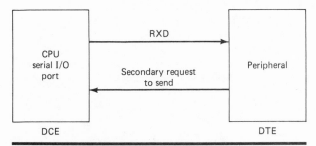

Fig. 4-7 Local Reverse Channel configuration.

signal from the data terminal equipment (DTE) to the data-communications equipment (DCE). One might think that the Reverse Channel should be implemented on the *Secondary Transmitted Data* signal line of the EIA standard since this is the signal line used to transmit data via the secondary channel from the DTE (printer) to the DCE (computer). However, the convention is that, when the secondary channel is used only for interrupting the flow of data on the primary channel, then the *Secondary Transmitted Data* signal is *not* normally used but rather the *Secondary Request to Send* when turned OFF is interpreted as an Interrupt condition. This makes sense because the only information that we need to transmit from printer to computer in our case is binary in nature:

OFF Stop sending data on the primary channel.

ON   Resume sending data on the primary channel.

Figure 4-7 illustrates a local Reverse Channel configuration involving peripheral equipment (destination) and a CPU (source).

If this method is used for speed-matching the slow peripheral with the fast computer, then the driver software must be able to recognize changes in state of pin 19, either as an Interrupt signal or by some other method.

Note that there is provision in the RS-232-C standard for Reverse Channel operation over a telecommunications channel. A remote Reverse Channel configuration is shown in Figure 4-8. The sequence of events for a remote Reverse Channel speed-matching operation proceeds as follows. After receiving transmitted data and falling behind, the peripheral equipment deactivates (turns off) the Secondary Request to Send line. Sensing this, the modem sends a logic 0 over the very low speed (10 baud or less—see Chapter 5) secondary frequency on the telephone channel. The modem at the CPU end senses this and turns off its RS-232-C Secondary Received Line Signal Detector signal. The Second-

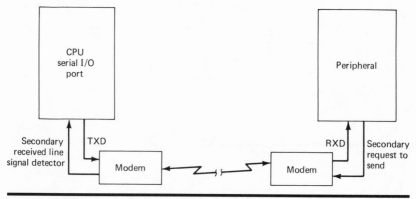

Fig. 4-8   Remote Reverse Channel configuration.

ary Received Line Signal Detector line can then be used to interrupt the CPU or raise a polled flag to tell the CPU to stop.

None of the Bell system modems implement the low-frequency secondary channel necessary to support remote Reverse Channel operation. For remote operation, by far the most common speed-matching techniques are XON/XOFF (first) and ETX/ACK (second). Even for local operation, microcomputer system interfaces rarely use the Reverse Channel technique. Thus, the dominant control mechanisms are the ones based upon the ASCII control characters.

COMMUNICATIONS LINE MONITORS

There are many good things that one can say about the benefits of data communications—it supports distributed processing, it allows remote users to access large computer facilities, and on and on. However, on the other side, when something goes wrong in a data communications environment, identifying the problem and finding the faulty element can be a nightmare. Even something as simple as getting a terminal to talk to a microcomputer CPU via an RS-232-C cable can sometimes be much more difficult than it sounds. As with any maintenance and repair function, the critical ingredient is often good diagnostic tools. The art and science of data communications troubleshooting and repair have progressed rapidly in the last decade. An industry has grown up with specialized companies dedicated to the design and development of instruments that implement functions to help analyze the performance of communications equipment.

Commercial Line Monitors

The commercial line monitors that are currently available are often microprocessor-based systems that perform a wide variety of functions. Some of these functions are discussed below.

On-line Data Recording

The data-recording function provided by communications line monitors permits a user to tap a channel physically and see what bits are flowing across it. The line monitor senses the logic levels on the channel and formats convenient displays in multiple number bases and/or codes for the user to analyze. In addition, most monitors provide extensive storage capacity so that data coming across can be saved and redisplayed at the user's option. The data storage capacity can vary from a few thousand bits to the multi-megabit range. Super large storage capacities are created by providing relatively high performance digital tape to unload the monitor's random access memory (RAM) in real time.

Event Traps

An event trap is a very useful mechanism that permits a user to specify a particular communications event as a stimulus for the line monitor to perform some predefined task. For example, if a special character or sequence of characters appears on the line, an event trap might be set to capture the byte sequences that were present immediately before and after the event. Most commercial line monitors can be programmed to trap several hundred bytes on either side of a predefined event. Other applications of traps are to use event occurrences for initiating and terminating RECORD mode and for sounding alarms. Event traps are not limited to character occurrences; they may also be events such as a certain bit pattern, a time-of-day trigger from a real-time clock, or the changing status of an EIA modem control lead. Thus certain external events can in some cases be used to enable and disable monitor functions automatically.

Protocol Support

Protocols typically supported by a single line monitor are asynchronous serial, binary synchronous communications (BISYNC), synchronous data-link control (SDLC), advanced digital-communications-control protocol (ADCCP), high-level data-link control (HDLC), digital data-communications message protocol (DDCMP), and certain variants. By supported, we mean that intelligence is built into the monitor to recognize protocol-unique events such as synchronization characters, error-checking characters, polls, acknowledgments, and negative acknowledgments.

Statistical Analysis Functions

Once event counts have been gathered and stored in real time, many line monitors support the application of statistical analysis to these event counts. Thus the monitor will support commands that find the mean, variance, standard deviation, and other statistics based upon observed event counts.

A line monitor of some variety is a necessity in any data-communications environment in which troubleshooting and diagnostic functions will be performed. However, commercially available line monitors, with capabilities similar to the ones we just described, are typically priced in the $4000 to $10,000 range. It is not feasible to use such a device for the occasional line monitoring that might be necessary for a single microcomputer configuration, but having some kind of monitoring capability could significantly simplify the solution of microcomputer-related data-communications problems.

Since most commercially available line monitors are implemented with microprocessors, it is reasonable to expect that a standard, general-purpose microcomputer system can perform many of the basic functions implemented by these more expensive specialized tools. In this section, we present a simple line monitor that we have implemented that is based upon a Z80 microcomputer.

A Simple Line Monitor

The Z80-based line monitor that we have developed consists of about 1 kbytes of software plus a communications channel tap. The tap is doc-

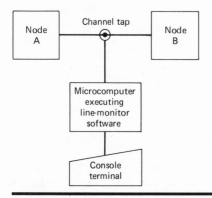

Fig. 4-9   Generic line monitor configuration.

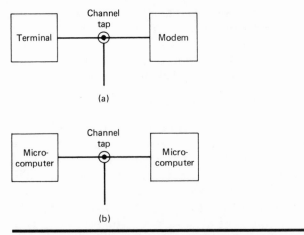

Fig. 4-10 *Two specific line monitor configurations:* (a) *channel tap between terminal and modem,* (b) *channel tap between two microcomputers.*

umented in Experiment 3-2 under the name *line monitor cable tap.* Via this tap, the user physically connects a Z80 microcomputer to a serial communications line. The software, via the tap, can read data traveling over the line, optionally display it in real time, store it, and display it later at the user's option. The particular signal levels on the communications line can be either TTL or RS-232-C—the tap can support both.

In the experiments at the end of this chapter, you will be able to use the line monitor to observe the flow of ASCII characters between two communicating devices. We have also included numerous exercises that are designed to provide you with some guidance, if you desire to upgrade the existing primitive functions of the line monitor.

Applications

Figure 4-9 shows an abstract picture of the type of configuration to which the line monitor is applicable. Figure 4-10 shows specific applications for the line monitor as currently implemented.

Using the line monitor, you will be able to observe what occurs on a 300 bps serial channel within a resolution of 1 bit time. Thus, for each ASCII character that passes through the line monitor's cable tab, each bit involved in that character transfer will be picked up and displayed. These bits include not only the seven logic levels that define the ASCII character, but also all of the asynchronous serial control bits—the start bit, the parity bit (if it exists), and the stop bit(s).

The next sections provide extensive documentation for the Z80-based line monitor. First we present a user's manual, then a discussion of the real-time display option, and finally a description of the hardware requirements for implementation of the line monitor. For the interested reader, subsequent sections give detailed information about the design and implementation of the line monitor software.

User's Manual

The line monitor is an interactive program that is operated with a menu-driven query-response format. In particular, after the initial start-up, the user is taken through a series of initialization questions that allow the software to format the output displays for the particular circumstances of the communications line to be monitored. Once initialization is complete, the user enters an operational mode in which three major line monitor functions are available. The following paragraphs identify and describe the initialization and operational functions implemented in the basic line monitor.

Initialization   The information that is provided by the user during line monitor initialization is specifically requested by three queries.

BIT COUNT: The first prompt from the line monitor requests the total bit count of the asynchronous serial characters to be monitored. This number must be entered in hexadecimal notation, using *capital* letters. This number counts the start bit, the data bits, the parity bit (if used), and the stop bits (1 or 2).

DATA BIT COUNT: The second prompt requests the total number of bits in the data portion of the characters being monitored. The choices are 6, 7, or 8.

REAL-TIME DISPLAY OPTION ENABLE AND DISABLE: The user may elect to enable or disable the real-time display. A continuous stream of about 4 or more data bytes traveling at 300 bps over the line being monitored can present a problem to a line monitor display terminal operating at 4800 bps or less. A subsequent section on the real-time display option explains the reasons for this difficulty in detail. It is enough to say here that a 4800 bps display terminal is too slow to support a real-time monitor display for continuous communications or for communications occurring in bursts of more than about four characters. Thus, if appropriate, the user may disable the real-time display. This has no effect upon the operational functions available to the user later. The only difference is that, during execution of the RECORD command, the monitored data bytes are not displayed as they come across the line. If the real-time display is enabled, the binary bit patterns of the incoming data are displayed on the console terminal as they are received.

Operation Once the initial parameters are specified, the line monitor software enters the operational mode. The two major functions implemented in our primitive line monitor are RECORD and PLAYBACK. Thus, the operational commands permit the user to control whether the software is collecting data from the communications channel or driving a display of observed bit patterns.

RECORD ON: This command causes the line monitor to begin reading and recording data as it flows through the cable tap connected to the communications channel. If the real-time display option is enabled, the observed bit patterns are displayed as they occur. If the line monitor display cannot support 9600 bps operation and there will be bursts of more than four characters transmitted on the line, then the real-time display option should be disabled. Otherwise, the real-time display provides the user with a dynamic picture of the communications events occurring on the line. Again, we refer you to a subsequent section for an explanation of why a 9600 bps display is required for real-time display.

Once the RECORD ON command is initiated, a short message will be displayed on the console indicating that the command is being executed:

RECORD is ON.

PLAYBACK: This command causes the bit stream that was recorded with the RECORD ON command to be played back (displayed on the console). For each character that the line monitor observes on the channel, a display is generated that gives the associated ASCII graphic character (if it exists), the hexadecimal representation of the data bits, and a binary sequence showing the start bit, the parity bit (if specified), and the stop bit(s).

For example, suppose the character C was transmitted across the communications line with 8 data bits, no parity, and 2 stop bits. Then the playback would display the following:

C 43 0110000101 1

The binary sequence is displayed in the following order:

Start	Data								(Parity)	Stop bit(s)	
0	1	1	0	0	0	0	1	0	—	1	1

with the data bits appearing in the following order:

Least significant					Most significant		
D_0	D_1	D_2	D_3	D_4	D_5	D_6	(D_7)

Once the PLAYBACK command has been initiated, a short message to that effect will be displayed on the console terminal:

PLAYBACK initiated.

CHANGE CONFIGURATION: The CHANGE CONFIGURATION command allows the user to change the initialization parameters by reexecuting the initialization sequence.

In addition to the above commands, there are two Break signals that are implemented by the line monitor.

PLAYBACK BREAK: Pressing any key during the PLAYBACK mode will suspend the display of the record buffer. Once a second key (*any key*) is pressed, the display of the record buffer will continue. The PLAYBACK Break function operates like a switch during the PLAYBACK mode to alternately stop and restart the display of the record buffer.

RECORD BREAK: Pressing any key during the RECORD mode will turn the RECORD mode OFF. Once the RECORD mode has been turned OFF, the user may restart the RECORD function by reissuing the RECORD ON command. Note that resuming the RECORD function during a transmission will not guarantee that the line monitor will be able to synchronize on a start bit. Thus, the resulting displays may produce erroneous characters. You will be able to tell that a synchronization problem exists by the presence of logic 0 "stop bits" in the bit sequence displays.

Once the RECORD function is stopped by using this Break feature, a short message to that effect is displayed on the console terminal.

RECORD is OFF.

Real-Time Display Option

If the console terminal has a 9600 bps display capability, then the data that are being captured off the communications line at the rate of 300 bps may be displayed in real time on the console. Each character is displayed in the following form:

.S.D0.D1.D2.D3.D4.D5.D6.D7.P.S1.S1.SP   or   CR.LF.

$$\begin{aligned}
\text{where } S &= \text{start bit} \\
D_0 \text{ through } D_7 &= \text{data bits} \\
P &= \text{parity bit} \\
S_1 \text{ and } S_2 &= \text{stop bits} \\
SP &= \text{Space} \\
CR &= \text{Carriage Return} \\
LF &= \text{Line Feed}
\end{aligned}$$

The above display format is for a maximal character specification: 1 start bit, 8 data bits, 1 parity bit, and 2 stop bits. Since the characters are displayed in rows of six characters, each separated by one space, every sixth character is followed by CR LF instead of SP.

We must address the real-time display format in this much detail to see why a 9600 bps display capability is required for the real-time display option. If the line monitor is tapped into a 300 bps line with 12 bit characters (such as the maximal one shown above) being transferred over it, this amounts to a requirement to display 25 characters per second in real time. For each of these 25 characters, the line monitor must produce a display like the one shown above. This display is made up of 13 characters most of the time, and 14 characters for every sixth character (at the end of a display line). If the line monitor display terminal is set up to accept 12-bit asynchronous serial characters, then for each second of real-time display operation, it must handle 25 $\times$ 13 $\times$ 12 bps plus 4 or 5 more CR characters. The total number of bits per second amounts to 3948 or 3960, depending upon whether 4 or 5 CRs are needed.

The total bit per second display throughput requirement that we have just computed is less than 4800 bps, so it is reasonable to think that a 4800 bps terminal can handle the real-time display operation. Later, when you perform the line monitor experiment at the end of this chapter, you might try setting up your display terminal for 4800 bps operation and enabling the real-time display option. You will find that it does not work.

How can we explain this? The reason is that 3948 or 3960 bps is perilously close to 4800 bps. In order to keep up with the incoming characters to display, the line monitor software must drive the display serial I/O port at almost top speed. A relatively small deviation from full-open throttle will cause the display to fall behind. The line monitor software, in reading the channel tap and in writing to the real-time display port, is operating in polled mode, alternately querying the tap port and display port. In particular, the line monitor must query the display port to see if it is ready to accept another display byte. A positive indication from the display port constitutes a *wet poll*, upon receipt of which the line monitor immediately outputs another byte for display. A *dry poll* is an indication to the line monitor that the display port is not ready for another byte. If the line monitor is driving the display port at top speed, the wait-for-poll time prior to a wet poll becomes highly significant. The wait-for-poll time is a function of the polling frequency associated with the display port, which, in turn, is a function of the processing performed between polls. As you can see, our analysis is taking us into a rather detailed study of the instructions that make up the line monitor software. This is very typical of data-communications software in particular and of real-time software in general.

We have taken you as far in this timing analysis as the law of diminishing returns will permit. It is clear what is happening. The wait-for-poll time is sufficiently great to reduce the effective bit rate of the display port to less than 3948 bps, which is what is required.

To add one more word about the real-time display problem, let us briefly see if we can reason why it works if the display port is configured for 9600 bps operation. At 9600 bps, the display port need only run at about half speed. Thus, the wait-for-poll time tolerance increases accordingly. The net effect is that the difference between demand and capacity for bit throughput is greater, thus providing tolerances that can be consistently met.

Since the real-time display option can be disabled, a 9600 bps display capability is not required to take advantage of the line monitor's RECORD and PLAYBACK functions.

Hardware Requirements

The line monitor, as implemented and documented for this book, is written in Z80 assembly language. Thus, as is, this implementation requires a Z80-based microcomputer system with about 16 kbytes of RAM. Actually, since the line monitor software is only about 1 kbytes, the monitor will operate on a system with as little as 3 or 4 kbytes of memory. However, the extra RAM is desirable for storage of recorded data.

In addition, the line monitor requires two ports—a serial console port and a parallel port. In order to be able to capture the raw bit pattern on an asynchronous serial line, a parallel port is used to capture the serial bit stream, including all control bits. Since the bit stream is serial, only *one* data bit on the parallel port is required. The line monitor software assumes that the serial bit stream is shifted in through bit D_0 of the parallel port. The particular parallel port that was used in our implementation has address C4. This address can be changed by appropriately altering one equate statement that defines the value for port A. The only pins of the parallel port that need to be implemented in the cable are the following:

- Data IN, bit D_0
- Signal Ground
- +5 V

The +5 V is needed to supply power to the cable-tap circuitry for converting RS-232-C levels to TTL levels appropriate for input over the parallel interface.

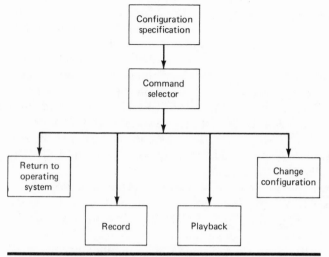

Fig. 4-11   High-level structure of the line monitor.

The channel tap is fully documented with complete instructions for building one in the experiments at the end of Chapter 3.

Line Monitor Design

The line monitor software consists of approximately 1 kbytes of Z80 code. A description of this software is given below. The high-level structure of the line monitor and a state-transition diagram that describes the functional flow of the software are discussed. The software modules that implement the RECORD and PLAYBACK functions are the most interesting, while the rest of the software is relatively straightforward. Thus we will limit our detailed discussion to these two modules. A complete listing of the Z80 assembly language source code for the line monitor is included in the experiments at the end of this chapter.

Logical Structure   The high-level structure of the line monitor is depicted in Figure 4-11. CONFIGURATION SELECTION is accomplished through the interactive initialization sequence in which the user specifies the total bits per character, the number of data bits per character, and enables or disables the real-time display option. The COMMAND SELECTION function presents a menu of available commands, namely:

- RECORD
- PLAYBACK
- CHANGE CONFIGURATION
- RETURN TO OPERATING SYSTEM

A state-transition diagram of the line monitor functions is given in Figure 4-12. It shows six possible states or modes associated with the line monitor. Note that the line monitor must be in precisely one mode at any one time. Transitions from state to state are accomplished by typing keys on the console terminal. For example, if the system is in the COMMAND SELECTION mode and a 2 is entered on the console terminal, then the system will leave the COMMAND SELECTION mode and enter the PLAYBACK mode. Similarly, if the system is in the PLAYBACK mode, no matter which key is pressed on the console terminal the system will enter the PAUSE mode. The subsequent reception of any key will reenable the PLAYBACK mode.

There are two completions that automatically trigger a mode change. Once the PLAYBACK mode is entered, the only mechanism by which one can return to the COMMAND SELECTION mode is to play back the entire buffer that was created during the previous RECORD mode session. Completion of the initialization sequence automatically causes the

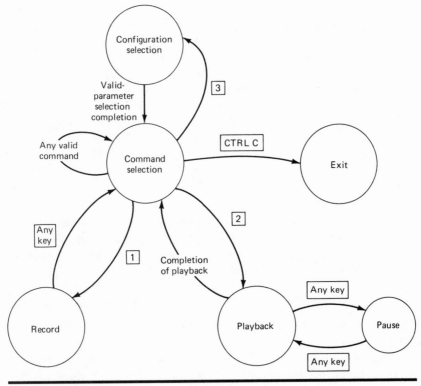

Fig. 4-12   Line monitor state-transition diagram.

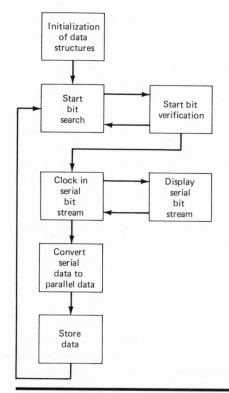

Fig. 4-13   RECORD *function flow diagram.*

line monitor to exit the CONFIGURATION SELECTION state and enter the COMMAND SELECTION state.

Implementation   The implementation of the two commands RETURN TO OPERATING SYSTEM and CHANGE CONFIGURATION is essentially trivial. Consequently, only the implementation of the RECORD and PLAYBACK commands will be discussed.

RECORD COMMAND IMPLEMENTATION: The implementation of the RECORD command is of interest particularly because it represents a software solution to the two major functions performed by universal asynchronous receiver-transmitters (UARTs), universal synchronous/asynchronous receiver-transmitters (USARTs), and serial input/output (SIO) type components namely parallel to serial conversion of data and bit timing for transmission and reception. A functional flow diagram of the implementation of the RECORD command is depicted in Figure 4-13 .

Once the initialization of the data structures is complete, a search is begun for a start bit. Since the transmission line idles in the high or marking state, a start bit is detected when a sampling of the line records a low or the presence of a logic 0. In order to ignore transients of a few microseconds' duration that might occur on the transmission line, a verification of the detection of the start bit is required. That is, after a delay of ½ bit time, the line is sampled again. If the line again is low, then this is taken as verification that a start bit exists on the transmission line. This is usually referred to as *spike detection*. Otherwise the search for a real start bit is resumed. As can be seen from Figure 4-13, if we continue to sample at 1-bit time intervals, we will be sampling each bit at approximately its midpoint. This is a desirable place to sample a bit time, since overshoots from the transition from the previous bit time should be minimized.

In this manner, sampling at the midpoint of each successive bit time continues until the required number of bits have been clocked into memory. As each bit is clocked in, it is converted to its ASCII representation, either 0 or 1, and it is displayed on the console terminal if the real-time display user option is enabled.

The serial bit stream of 9 to 12 bits for each character is subsequently rotated into 2 bytes of memory and stored there for later access, if required. It is this process of rotating or shifting the serial bit stream into a parallel form that represents the serial to parallel conversion that must take place. Thus, this function of the line monitor represents a software implementation of a shift register.

COMMAND IMPLEMENTATION: A functional flow diagram for the PLAY-BACK command implementation is depicted in Figure 4-14.

Once the initialization for the user's current display format is complete, the bit string for one character is retrieved from memory and rotated to isolate the data bits from the control bits. The data bits are then displayed in their ASCII representation and ASCII to hexadecimal conversion is performed to display the hexadecimal representation of the data.

Note that the line monitor software calls an external routine named BINH2 to perform this conversion. BINH2 is a subroutine that is supplied by the Cromenco CDOS operating system that we happened to be using. You may also have a system-supplied binary to hexadecimal conversion routine that you can use via a system call. If not, the conversion function is very simple to implement. Essentially you have to isolate the two nybbles that comprise the byte that you wish to display in hexadecimal notation. Then add each of these nybbles to the hex 30. The result is the ASCII code for the specific byte's hexadecimal representation. By outputting the two sums to an ASCII terminal, most significant nybble first, you can display the hexadecimal representation for a byte.

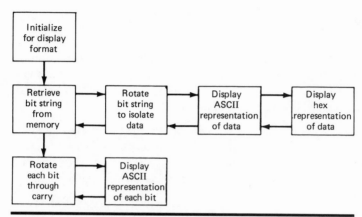

Fig. 4-14 **PLAYBACK** *function flow diagram.*

Finally, since the PLAYBACK mode also displays each of the control and data bits associated with the character, it is necessary to "serialize" the data again and rotate each bit through the carry flag to determine and display its ASCII representation.

Data Structures   The most significant data structures of the line monitor are identified and discussed in this section.

BITS: This is a 1-byte variable that contains the total number of bits per character that are being transmitted on the serial link that is being monitored. This total includes all data bits together with all control bits. The variable obtains its value during the CONFIGURATION SELECTION or initialization session with the user.

DATA: This variable represents the total number of data bits per character that are being transmitted on the monitored link. The value of DATA is determined in the same manner and at the same time as the variable BITS.

STRING: The output of the binary to hexadecimal conversion routine leaves the hexadecimal representation of the binary data in this variable. This conversion was accomplished by using an existing routine BINH2 available in the Cromemco Disk Operating System (CDOS) assembler library, ASMLIB. The development of a routine that will perform the same functions as BINH2 is discussed above.

COUNT: This is a variable that controls the format of the real-time display. The value of COUNT represents the number of character bit strings that are displayed in one line of the real-time display. This value is assigned at the EQUATE statement at the beginning of the program.

RTD: This variable is determined at the same time and in the same manner as the values of BITS and DATA are determined. It represents a flag with the property that if its value is 1, the real-time display is enabled, and if its value is 0, the real-time display is disabled.

INPTR: This variable contains the address of the next available location for the storage of data in the input buffer. Other variables associated with the input buffer are discussed below. Initially, INPTR has the value INBUFF, which is the lowest memory location associated with the input buffer.

BUFTOP: Once the RECORD mode is disabled, the value of the pointer INPTR is stored in the variable BUFTOP. It represents the top of the input buffer that was created by the previous execution of the RECORD function.

INBUFF: This represents the initial storage location of the input buffer. There is no end-of-buffer test or warning associated with the use of the input buffer. The size of the buffer is simply determined by the memory space available. As an enhancement to the line monitor software, some version of buffer management should be implemented, so that a buffer overflow that overwrites the executing code of the line monitor software cannot occur.

EXPERIMENTS

The following set of experiments is designed to integrate what you have read in the first four chapters of this book. In Chapter 2 you read about polled and interrupt-driven I/O. Also, in Chapter 2 you read about the asynchronous serial protocol and the methods that it uses for character synchronization. In Chapter 3, you read about RS-232-C and fabricated cables and a hardware line monitor (channel tap) that can be used to observe bits being transferred over a cable. In this chapter, the first section dealt with the ASCII encoding scheme and the next section discussed the problem of speed matching or flow control. All of these issues are illustrated in the following set of experiments.

The first experiment shows you how to use the line monitor software and channel tap to observe asynchronous serial ASCII characters as streams of bits transferred across a 300 bps communications link. The subsequent experiments investigate the significance of timing in data output operations by showing you a series of assembly language programs, each of which presents its own technique for resolving a particular timing issue. The final experiments show you how to implement interrupt-driven I/O, first using an S100-based serial I/O board equipped with hardware to support interrupt generation, and second using a TRS-80 microcomputer system with a special interrupt circuit that is easy and inexpensive to build.

Experiment 4-1: A Simple Line Monitor

Purpose

This experiment is designed to help you become more familiar with the link protocol used with asynchronous serial communications. This experiment will use the line monitor software and channel tap that were fabricated in an experiment in Chapter 3. With this equipment you will be able to display the bit patterns that are transmitted during an asynchronous data transfer.

Equipment

We have implemented the line monitor software in Z80 microprocessor assembly language. Thus, this experiment requires a Z80-based microcomputer with a terminal (for display of monitored data), a parallel I/O port (for the channel tap), and at least 4 kbytes of RAM. The terminal used to display data recorded by the line monitor will be referred to as the *monitor console* or the *monitor display terminal.*

The channel tap fabricated in Experiment 3-2 is also required. A second terminal, or source of serially transmitted data, is also required for this experiment. This serial data source will be referred to as the *test terminal* throughout this experiment. We hope that, in this fashion, we can always be clear as to exactly which terminal we are talking about, the monitor display console or the test terminal.

Figure 4-15 on page 192 shows how the equipment components described above will be configured for this experiment.

Software

Listing 4-1 is the full Z80 assembly language source code listing for the line monitor.

Step 1

Connect the equipment for this experiment as shown in Figure 4-15 . In this experiment, the characters transmitted over the communications link that you will be monitoring will not come in bursts of more than one character at a time since the characters are keyed manually. Thus, you should be able to operate the line monitor with the real-time display option enabled at 300 baud.

LISTING 4-1: *Source Code for a Simple Z80-Based Line Monitor*

```
                    0001    ;
                    0002    ;LINE MONITOR
                    0003    ;VERSION 1.0 FEBRUARY 14, 1981
                    0004    ;COPYRIGHT 1981 NICHOLS,NICHOLS AND MUSSON
                    0005    ;
                    0006    ;
                    0007            EXT BINH2
        (007B)      0008    HBAUD   EQU     07BH
        (00F6)      0009    BAUD    EQU     0F6H
        (00C4)      0010    PORTA   EQU     0C4H
        (000D)      0011    CR      EQU     0DH
        (000A)      0012    LF      EQU     0AH
        (0020)      0013    SPACE   EQU     20H
        (0019)      0014    EM      EQU     19H
        (0000)      0015    TBASE   EQU     00H
        (0001)      0016    TDATA   EQU     01H
        (0012)      0017    OFF     EQU     12H
        (0003)      0018    CTRLC   EQU     03H
        (0001)      0019    RTDOFF  EQU     01H
        (0005)      0020    CDOS    EQU     05H
        (0096)      0021    MOTOR   EQU     96H
                    0022    ;
                    0023    ;
0000' 0E96          0024    LM:     LD C,96H        ;TURN OFF DISK MOTOR
0002' CD0500        0025            CALL CDOS
0005' 0E00          0026    BCPMT:  LD C,TBASE
0007' 213602'       0027            LD HL,BITMSG    ;PROMPT FOR BIT COUNT
000A' CD2402'       0028            CALL TMSG
000D' 0E00          0029            LD C,TBASE
000F' CDDE01'       0030            CALL GETC       ;GET BIT COUNT
0012' CDD201'       0031            CALL PUTC       ;ECHO BIT COUNT
0015' D637          0032            SUB 37H         ;ASJUST FOR ASCII
0017' FE02          0033            CP 2H           ;TREAT 9 BITS SPECIAL
0019' 2805          0034            JR Z,NINE
001B' 323F04'       0035            LD (BITS),A     ;STORE BIT COUNT
001E' 1805          0036            JR DBPMT
0020' 3E09          0037    NINE:   LD A,09H        ;LAST CASE
0022' 323F04'       0038            LD (BITS),A
0025' 0E00          0039    DBPMT:  LD C,TBASE
0027' 216E02'       0040            LD HL,DATMSG    ;PROMPT FOR DATA BITS ?
002A' CD2402'       0041            CALL TMSG
002D' 0E00          0042            LD C,TBASE
002F' CDDE01'       0043            CALL GETC       ;GET # DATA BITS
0032' CDD201'       0044            CALL PUTC       ;ECHO TO TERMINAL
0035' E67F          0045            AND 7FH         ;MAKE IT ASCII
0037' FE36          0046            CP 36H          ;6 DATA BITS PER CHAR
0039' 280A          0047            JR Z,DATA6
003B' FE37          0048            CP 37H          ;7 DATA BITS PER CHAR
003D' 280A          0049            JR Z,DATA7
003F' FE38          0050            CP 38H          ;8 DATA BITS PER CHAR
0041' 280A          0051            JR Z,DATA8
0043' 18BB          0052            JR LM           ;ELSE START OVER
0045' 3E3F          0053    DATA6:  LD A,3FH        ;MASK BYTE FOR 6
0047' 1806          0054            JR REMEM
0049' 3E7F          0055    DATA7:  LD A,7FH        ;MASK BYTE FOR 7
004B' 1802          0056            JR REMEM
004D' 3EFF          0057    DATA8:  LD A,0FFH       ;MASK FOR 8 BITS
004F' 324004'       0058    REMEM:  LD (DATA),A     ;REMEMBER THIS MASK
0052' 0E00          0059    RTDPMT: LD C,TBASE
0054' 219E02'       0060            LD HL,RTDMSG    ;PROMPT FOR RTD CHOICE
0057' CD2402'       0061            CALL TMSG
005A' 0E00          0062            LD C,TBASE      ;GET CHOICE
005C' CDDE01'       0063            CALL GETC
005F' 324404'       0064            LD (RTD),A      ;RTD = (COUNT + 1)
0062' CDD201'       0065            CALL PUTC       ;ECHO CHOICE
0065' 0E00          0066    COMPMT: LD C,TBASE
0067' 21F102'       0067            LD HL,PROMSG    ;PROMPT MESSAGE
006A' CD2402'       0068            CALL TMSG
006D' 0E00          0069            LD C,TBASE
006F' CDDE01'       0070            CALL GETC       ;GET COMMAND
0072' FE31          0071            CP 31H
```

LISTING 4-1: Source Code for a Simple Z80-Based Line Monitor (Continued)

```
0074'  280F        0072              JR   Z,RECORD     ;TURN RECORD ON
0076'  FE32        0073              CP   32H
0078'  282E        0074              JR   Z,PLAY       ;PLAY BACK MODE
007A'  FE33        0075              CP   33H
007C'  2882        0076              JR   Z,LM         ;CHANGE CONFIGURATION
007E'  FE03        0077              CP   CTRLC
0080'  CA0000      0078              JP   Z,0000H      ;RETURN TO OPERATING SYSTEM
0083'  18E0        0079              JR   COMPMT       ;ALL OTHER COMMANDS DEFAULT
                   0080              ;
0085'  0E00        0081    RECORD:   LD   C,TBASE      ;RECORD MODE ON
0087'  210B04'     0082              LD   HL,MSG1      ;MESSAGE TO CONSOLE
008A'  CD2402'     0083              CALL TMSG
008D'  215F04'     0084              LD   HL,INBUFF
0090'  224504'     0085              LD   (INPTR),HL   ;RESET BUFFER POINTER
0093'  C34A01'     0086              JP   MONTOR
                   0087              ;
0096'  DB01        0088    STOP:     IN   A,(TDATA)    ;DUMMY TO DUMP CHAR
0098'  0E00        0089              LD   C,TBASE      ;RECORD OFF MESSAGE
009A'  211A04'     0090              LD   HL,MSG2
009D'  CD2402'     0091              CALL TMSG
00A0'  2A4504'     0092              LD   HL,(INPTR)   ;STORE BUFFER POINTER
00A3'  224704'     0093              LD   (BUFTOP),HL
00A6'  18BD        0094              JR   COMPMT       ;COMMAND MODE
                   0095              ;
                   0096              ;
00A8'  0E00        0097    PLAY:     LD   C,TBASE      ;PLAYBACK MODE
00AA'  212A04'     0098              LD   HL,MSG3
00AD'  CD2402'     0099              CALL TMSG         ;MESSAGE TO CONSOLE
00B0'  37          0100              SCF
00B1'  3F          0101              CCF               ;CLEAR CARRY FLAG
00B2'  2A4704'     0102              LD   HL,(BUFTOP)
00B5'  015F04'     0103              LD   BC,INBUFF
00B8'  ED42        0104              SBC  HL,BC        ;COMPUTE HL=HL-INBUFF
00BA'  FD215F04'   0105              LD   IY,INBUFF    ;POINT TO INPUT BUFFER
00BE'  7D          0106              LD   A,L          ;CKECK FOR EMPTY BUFFER
00BF'  B4          0107              OR   H
00C0'  28A3        0108              JR   Z,COMPMT     ;START OVER IF EMPTY
                   0109              ;
00C2'  DD214904'   0110    ACROSS:   LD   IX,SHELF     ;SET UP FOR ONE LINE
00C6'  DD361204    0111              LD   (IX+OFF),04H ;ACROSS DISPLAY
00CA'  DD214904'   0112    BEGIN:    LD   IX,SHELF     ;POINT TO TEMPORARY BUFFER
00CE'  3A3F04'     0113              LD   A,(BITS)     ;#BITS PER BYTE
00D1'  47          0114              LD   B,A
00D2'  FD5E00      0115              LD   E,(IY)       ;GET MSB FROM BUFFER
00D5'  FD23        0116              INC  IY
00D7'  FD5600      0117              LD   D,(IY)       ;GET LSB FROM BUFFER
00DA'  FD23        0118              INC  IY
                   0119              ;
00DC'  CDEB01'     0120              CALL LMBE         ;LINE MONITOR BACK END
                   0121              ;
00DF'  CB13        0122    ROTATE:   RL   E
00E1'  CB12        0123              RL   D            ;ROTATE BITS TO CARRY
00E3'  3806        0124              JR   C,STONE      ;STORE AN ASCII ONE
00E5'  DD360030    0125              LD   (IX),30H     ;STORE AN ASCII ZERO
00E9'  1804        0126              JR   NEXT
00EB'  DD360031    0127    STONE:    LD   (IX),31H
00EF'  DD23        0128    NEXT:     INC  IX
00F1'  10EC        0129              DJNZ ROTATE       ;DO IT FOR ALL BITS
                   0130              ;
00F3'  3A3F04'     0131              LD   A,(BITS)     ;PREPARE FOR OUTPUT
00F6'  47          0132              LD   B,A
00F7'  0E00        0133              LD   C,TBASE
00F9'  DD2B        0134              DEC  IX
00FB'  DD7E00      0135    UNWIND:   LD   A,(IX)
00FE'  CDD201'     0136              CALL PUTC
0101'  DD2B        0137              DEC  IX
0103'  10F6        0138              DJNZ UNWIND       ;OUTPUT ALL BITS
0105'  DD214904'   0139              LD   IX,SHELF
0109'  DD3512      0140              DEC  (IX+OFF)     ;COUNTER FOR ONE LINE
010C'  280F        0141              JR   Z,NXTLIN
010E'  3E20        0142              LD   A,SPACE
```

LISTING 4-1: Source Code for a Simple Z80-Based Line Monitor (Continued)

```
0110'  0E00        0143              LD   C,TBASE
0112'  CDD201'     0144              CALL PUTC        ;SPACE BETWEEN ASCII
0115'  CDD201'     0145              CALL PUTC        ;REPRESENTATION OF
0118'  CDD201'     0146              CALL PUTC        ;EACH BIT
011B'  1814        0147              JR   PCHECK
011D'  0E00        0148      NXTLIN: LD   C,TBASE     ;OUTPUT CR,LF FOR
011F'  3E0D        0149              LD   A,CR        ;LINEFOLD CONTROL
0121'  CDD201'     0150              CALL PUTC
0124'  3E0A        0151              LD   A,LF
0126'  CDD201'     0152              CALL PUTC
0129'  DD214904'   0153              LD   IX,SHELF
012D'  DD361204    0154              LD   (IX+OFF),04H
0131'  DB00        0155      PCHECK: IN   A,(TBASE)   ;PAUSE CHECK
0133'  17          0156              RLA              ;FOR ANY CHARACTER FROM
0134'  17          0157              RLA              ;CONSOLE
0135'  300A        0158              JR   NC,FWD
0137'  DB01        0159              IN   A,(TDATA)   ;DUMMY INPUT TO RESET RDA
0139'  DB00        0160      PAUSE:  IN   A,(TBASE)   ;PAUSE UNTIL A SECOND
013B'  17          0161              RLA              ;CHARACTER ARRIVES FROM
013C'  17          0162              RLA              ;CONSOLE
013D'  30FA        0163              JR   NC,PAUSE
013F'  DB01        0164              IN   A,(TDATA)   ;DUMMY INPUT TO RESET RDA
0141'  2B          0165      FWD:    DEC  HL
0142'  2B          0166              DEC  HL          ;UPDATE COUNTER
0143'  7D          0167              LD   A,L
0144'  B4          0168              OR   H           ;IS COUNTER 0?
0145'  2083        0169              JR   NZ,BEGIN
0147'  C36500'     0170              JP   COMPMT
                   0171      ;
                   0172      ;
014A'  FD214304'   0173      MONTOR: LD   IY,COUNT    ;MONITOR THE LINE
014E'  FD360006    0174              LD   (IY),06H    ;6 CHARACTERS ACROSS
0152'  16FF        0175      LINE:   LD   D,0FFH
0154'  1EFF        0176              LD   E,0FFH
0156'  3A3F04'     0177              LD   A,(BITS)    ;INITIALIZE BIT COUNTER
0159'  47          0178              LD   B,A
015A'  DBC4        0179      LOOK:   IN   A,(PORTA)   ;PARALLEL PORT A
015C'  1F          0180              RRA              ;LOOK FOR MONTOR BIT
015D'  3009        0181              JR   NC,GETIT
015F'  DB00        0182      CHECKT: IN   A,(TBASE)   ;CHECK TERMINAL
0161'  17          0183              RLA
0162'  17          0184              RLA              ;CHECK RDA FOR TERMINAL
0163'  DA9600'     0185              JP   C,STOP
0166'  18F2        0186              JR   LOOK        ;CHECK LINE AGAIN
                   0187              ;
0168'  C5          0188      GETIT:  PUSH BC          ;SAVE STATE
0169'  017B00      0189              LD   BC,HBAUD    ;1/2 BIT TIME
016C'  CDCA01'     0190              CALL DELAY       ;WAIT IT OUT
016F'  C1          0191              POP  BC          ;RESTORE STATE
0170'  DBC4        0192              IN   A,(PORTA)   ;CHECK PORT AGAIN
0172'  1F          0193              RRA              ;FOR MONTOR BIT
0173'  38E5        0194              JR   C,LOOK      ;IF NOT LOOK AGAIN
0175'  DBC4        0195      GETBIT: IN   A,(PORTA)   ;READ ONE BYTE
0177'  1F          0196              RRA              ;ROTATE BIT 0 TO CARRY
0178'  3804        0197              JR   C,PONE
017A'  3E30        0198              LD   A,30H       ;OUTPUT ASCII 0 PREP
017C'  1802        0199              JR   TERM
017E'  3E31        0200      PONE:   LD   A,31H       ;OUTPUT ASCII 1 PREP
0180'  0E00        0201      TERM:   LD   C,TBASE
0182'  FDCB0146    0202              BIT  0,(IY+RTDOFF) ;REAL TIME MODE ON?
0186'  C4D201'     0203              CALL NZ,PUTC
0189'  CB1A        0204              RR   D           ;SAVE BIT IN DE
018B'  CB1B        0205              RR   E           ;LSB'S TO E
018D'  05          0206              DEC  B
018E'  280A        0207              JR   Z,STORE     ;STORE THIS CHAR
0190'  C5          0208              PUSH BC          ;SAVE STATE
0191'  01F600      0209              LD   BC,BAUD     ;1 BIT TIME LESS CODE
0194'  CDCA01'     0210              CALL DELAY
0197'  C1          0211              POP  BC          ;RESTORE STATE
0198'  18DB        0212              JR   GETBIT      ;GET NEXT BIT
                   0213              ;
```

LISTING 4-1: Source Code for a Simple Z80-Based Line Monitor (Continued)

```
019A'  2A4504'    0214    STORE:   LD HL,(INPTR)       ;POINT TO IN-BUFFER
019D'  73          0215             LD (HL),E           ;MOVE LSB'S TO BUFF
019E'  23          0216             INC HL              ;UPDATE POINTER
019F'  72          0217             LD (HL),D           ;MOVE MSB'S TO BUFF
01A0'  23          0218             INC HL              ;UPDATE POINTER
01A1'  224504'    0219             LD (INPTR),HL       ;SAVE POINTER
01A4'  0E00        0220             LD C,TBASE
01A6'  3E20        0221             LD A,SPACE
01A8'  FDCB0146    0222             BIT 0,(IY+RTDOFF)   ;REAL TIME DISPLAY ?
01AC'  C4D201'    0223             CALL NZ,PUTC
01AF'  FD3500      0224             DEC (IY)
01B2'  209E        0225             JR NZ,LINE
01B4'  0E00        0226             LD C,TBASE          ;OUTPUT CR,LF AT
01B6'  3E0D        0227             LD A,CR             ;PROPER TIME
01B8'  FDCB0146    022              BIT 0,(IY+RTDOFF)
01BC'  C4D201'    0229             CALL NZ,PUTC
01BF'  3E0A        0230             LD A,LF
01C1'  FDCB0146    0231             BIT 0,(IY+RTDOFF)
01C5'  C4D201'    0232             CALL NZ,PUTC
01C8'  1880        0233             JR MONTOR
                    0234             ;
                    0235             ;
                    0236             NAME DELAY
                    0237    ;
01CA'  F5          0238    DELAY:   PUSH AF
01CB'  0B          0239    STALL:   DEC BC              ;BC CONTAINS DELAY COUNT
01CC'  78          0240             LD A,B              ;IS BC=0000H
01CD'  B1          0241             OR C
01CE'  20FB        0242             JR NZ,STALL         ;IF NOT STALL MORE
01D0'  F1          0243             POP AF
01D1'  C9          0244             RET
                    0245    ;
                    0246    ;
                    0247             NAME PUTC
                    0248             ;C MUST CONTAIN BASE PORT ADDRESS
                    0249             ;OUT PUTS CHARACTER IN A
01D2'  F5          0250    PUTC:    PUSH AF             ;SAVE CHARACTER
01D3'  ED78        0251    LOOP:    IN A,(C)            ;CHECK TRANSMITTER
01D5'  17          0252             RLA                 ;BUFFER EMPTY FLAG
01D6'  30FB        0253             JR NC,LOOP
01D8'  0C          0254             INC C               ;CHANGE PORT TO DATA
01D9'  F1          0255             POP AF              ;RESTORE CHARACTER
01DA'  ED79        0256             OUT (C),A           ;OUTPUT CHARACTER
01DC'  0D          0257             DEC C               ;RESTORE C
01DD'  C9          0258             RET
                    0259             ;
                    0260             ;GETC GETS ONE CHARACTER
                    0261             ;C MUST BE BASE PORT ADDRESS
                    0262             ;RETURNS CHARACTER IN A
01DE'  ED78        0263    GETC:    IN A,(C)            ;CHECK RECEIVE DATA
01E0'  17          0264             RLA                 ;AVAILABLE FLAG
01E1'  17          0265             RLA
01E2'  30FA        0266             JR NC,GETC
01E4'  0C          0267             INC C               ;CHANGE PORT TO DATA
01E5'  ED78        0268             IN A,(C)            ;INPUT CHARACTER
01E7'  CBBF        0269             RES 7,A             ;MAKE IT ASCII
01E9'  0D          0270             DEC C               ;RESTORE C
01EA'  C9          0271             RET
                    0272             ;
                    0273             ;
                    0274             ;
01EB'  E5          0275    LMBE:    PUSH HL             ;LINE MONITOR BACK
01EC'  D5          0276             PUSH DE             ;END
01ED'  C5          0277             PUSH BC
01EE'  214104'    0278             LD HL,STRING
01F1'  7A          0279             LD A,D              ;GET MOST SIG BITS
01F2'  43          0280             LD B,E              ;GET LEAST SIG BITS
01F3'  1F          0281    SHIFT:   RRA                 ;ROTATE A INTO CARRY
01F4'  CB18        0282             RR B                ;ROTATE B INTO CARRY
01F6'  38FB        0283             JR C,SHIFT          ;LOOK FOR START BIT
01F8'  3A4004'    0284             LD A,(DATA)         ;DATA BIT MASK BYTE
```

LISTING 4-1: Source Code for a Simple Z80-Based Line Monitor (Continued)

```
01FB'  A0           0285                  AND B              ;MASK HIGH ORDER ONES
01FC'  0E00         0286                  LD C,TBASE
01FE'  CDD201'      0287                  CALL PUTC          ;OUTPUT ASCII FORM
0201'  CD0000#      0288                  CALL BINH2         ;GET HEX FORM
0204'  3E20         0289                  LD A,SPACE         ;OUTPUT SPACE FOR
0206'  0E00         0290                  LD C,TBASE         ;FORMAT CONTROL
0208'  CDD201'      0291                  CALL PUTC
020B'  214104'      0292                  LD HL,STRING       ;BINH2 LEFT ANSWER
020E'  7E           0293                  LD A,(HL)          ;POINTER
020F'  0E00         0294                  LD C,TBASE
0211'  CDD201'      0295                  CALL PUTC          ;FIRST HEX CHARACTER
0214'  23           0296                  INC HL
0215'  7E           0297                  LD A,(HL)
0216'  CDD201'      0298                  CALL PUTC          ;SECOND HEX CHARACTER
0219'  3E20         0299                  LD A,SPACE
021B'  0E00         0300                  LD C,TBASE
021D'  CDD201'      0301                  CALL PUTC          ;OUTPUT SPACE FOR
0220'  C1           0302                  POP BC             ;FORMAT CONTROL
0221'  D1           0303                  POP DE
0222'  E1           0304                  POP HL
0223'  C9           0305                  RET
                    0306                  ;
0224'  E5           0307    TMSG:         PUSH HL            ;TRANSMIT MESSAGE TO
0225'  C5           0308                  PUSH BC            ;CONSOLE
0226'  F5           0309                  PUSH AF
                    0310                  ;
0227'  7E           0311    TMAIL:        LD A,(HL)          ;HL POINTS TO MESSAGE
0228'  FE19         0312                  CP EM              ;END OF MESSAGE MARK
022A'  2806         0313                  JR Z,BACK
022C'  CDD201'      0314                  CALL PUTC          ;OUTPUT CHARACTER
022F'  23           0315                  INC HL             ;UPDATE POINTER
0230'  18F5         0316                  JR TMAIL           ;REPEAT UNTIL EM
0232'  F1           0317    BACK:         POP AF
0233'  C1           0318                  POP BC
0234'  E1           0319                  POP HL
0235'  C9           0320                  RET
                    0321                  ;
                    0322                  ;
0236'  0D0A0D0A     0323    BITMSG: DB    CR,LF,CR,LF
023A'  456E7465     0324            DB    'Enter BIT COUNT per character'
0257'  20696E20     0325            DB    ' in HEX: 9,A,B,or C',CR,LF
026C'  3E19         0326            DB    '>',EM
026E'  0D0A0D0A     0327    DATMSG: DB    CR,LF,CR,LF
0272'  456E7465     0328            DB    'Enter DATA BITS per character:'
0290'  20362C37     0329            DB    ' 6,7, OR 8',CR,LF,'>',EM
029E'  0D0A0D0A     0330    RTDMSG: DB    CR,LF,CR,LF
02A2'  456E7465     0331            DB    'Enter 1 to ENABLE real time'
02BD'  20646973     0332            DB    ' display:',CR,LF
02C8'  456E7465     0333            DB    'Enter 0 to DISABLE real time'
02E4'  20646973     0334            DB    ' display:',CR,LF,'>',EM
                    0335            ;
02F1'  0D0A0A20     0336    PROMSG: DB    CR,LF,LF,20H,20H,20H,20H,20H,20H
02FA'  20202020     0337            DB    20H,20H,20H,20H
02FE'  4D6F6E69     0338            DB    'Monitor Commands',CR,LF
0310'  0A2A2A2A     0339            DB    LF,'****************************'
032D'  2A2A2A2A     0340            DB    '******************'
033E'  0D0A0A31     0341            DB    CR,LF,LF,'1 - RECORD ON',CR,LF
0350'  32202D20     0342            DB    '2 - PLAYBACK',CR,LF
035E'  33202D20     0343            DB    '3 - CONFIGURATION CHANGE',CR,LF
0378'  43544C20     0344            DB    'CTL "C" - RETURN TO CDOS',CR,LF
0392'  0A0A0A41     0345            DB    LF,LF,LF,'Any Key-PAUSE DURING PLAY'
03AE'  206F7220     0346            DB    ' or RECORD OFF',CR,LF
03BE'  416E7920     0347            DB    'Any Key-RESUME AFTER PAUSE'
03D8'  0D0A0A2A     0348            DB    CR,LF,LF,'****************************'
03F1'  2A2A2A2A     0349            DB    '********************'
0408'  0D0A19       0350            DB    CR,LF,EM
040B'  5245434F     0351    MSG1:   DB    'RECORD is on',CR,LF,EM
041A'  5245434F     0352    MSG2:   DB    'RECORD is off',CR,LF,EM
042A'  504C4159     0353    MSG3:   DB    'PLAYBACK Initiated',CR,LF,EM
043F'  0A           0354    BITS:   DB    0AH
0440'  FF           0355    DATA:   DB    0FFH
```

LISTING 4-1: Source Code for a Simple Z80-Based Line Monitor (Continued)

```
0441' 0000        0356  STRING: DW      0000H
                  0357          ;
                  0358  ;
0443' 00          0359  COUNT:  DB      00H
0444' 00          0360  RTD:    DB      00H
0445' 5F04'       0361  INPTR:  DW      INBUFF
0447' 5F04'       0362  BUFTOP: DW      INBUFF
0449' (0016)      0363  SHELF:  DS      16H
045F' (0400)      0364  INBUFF: DS      400H
                  0365  ;
                  0366  ;
085F' (0000')     0367          END LM
```

Program Length 085F (2143)

```
Symbol     Value   Defn   References

ACROSS     00C2'   0110
BACK       0232'   0317   0313
BAUD       00F6    0009   0209
BCPMT      0005'   0026
BEGIN      00CA'   0112   0169
BINH2    X 0000#   0007   0288
BITMSG     0236'   0323   0027
BITS       043F'   0354   0035 0038 0113 0131 0177
BUFTOP     0447'   0362   0093 0102
CDOS       0005    0020   0025
CHECKT     015F'   0182
COMPMT     0065'   0066   0079 0094 0108 0170
COUNT      0443'   0359   0173
CR         000D    0011   0149 0227 0323 0323 0325 0327 0327 0329 0330 0330 0332
                          0334 0336 0338 0341 0341 0342 0343 0344 0346 0348 0350
                          0351 0352 0353
CTRLC      0003    0018   0077
DATA       0440'   0355   0058 0284
DATA6      0045'   0053   0047
DATA7      0049'   0055   0049
DATA8      004D'   0057   0051
DATMSG     026E'   0327   0040
DBPMT      0025'   0039   0036
DELAY      01CA'   0238   0190 0210
EM         0019    0014   0312 0326 0329 0334 0350 0351 0352 0353
FWD        0141'   0165   0158
GETBIT     0175'   0195   0212
GETC       01DE'   0263   0030 0043 0063 0070 0266
GETIT      0168'   0188   0181
HBAUD      007B    0008   0189
INBUFF     045F'   0364   0084 0103 0105 0361 0362
INPTR      0445'   0361   0085 0092 0214 0219
LF         000A    0012   0151 0230 0323 0323 0325 0327 0327 0329 0330 0330 0332
                          0334 0336 0336 0338 0339 0341 0341 0341 0342 0343 0344
                          0345 0345 0345 0346 0348 0348 0350 0351 0352 0353
LINE       0152'   0175   0225
LM         0000'   0024   0052 0076 0367
LMBE       01EB'   0275   0120
LOOK       015A'   0179   0186 0194
LOOP       01D3'   0251   0253
MONTOR     014A'   0173   0086 0233
MOTOR      0096    0021
MSG1       040B'   0351   0082
MSG2       041A'   0352   0090
MSG3       042A'   0353   0098
NEXT       00EF'   0128   0126
NINE       0020'   0037   0034
NXTLIN     011D'   0148   0141
OFF        0012    0017   0111 0140 0154
PAUSE      0139'   0160   0163
PCHECK     0131'   0155   0147
PLAY       00A8'   0097   0074
```

LISTING 4-1: Source Code for a Simple Z80-Based Line Monitor (Continued)

```
PONE      017E'   0200   0197
PORTA     00C4    0010   0179 0192 0195
PROMSG    02F1'   0336   0067
PUTC      01D2'   0250   0031 0044 0065 0136 0144 0145 0146 0150 0152 0203 0223
                         0229 0232 0287 0291 0295 0298 0301 0314
RECORD    0085'   0081   0072
REMEM     004F'   0058   0054 0056
ROTATE    00DF'   0122   0129
RTD       0444'   0360   0064
RTDMSG    029E'   0330   0060
RTDOFF    0001    0019   0202 0222 0228 0231
RTDPMT    0052'   0059
SHELF     0449'   0363   0110 0112 0139 0153
SHIFT     01F3'   0281   0283
SPACE     0020    0013   0142 0221 0289 0299
STALL     01CB'   0239   0242
STONE     00EB'   0127   0124
STOP      0096'   0088   0185
STORE     019A'   0214   0207
STRING    0441'   0356   0278 0292
TBASE     0000    0015   0026 0029 0039 0042 0059 0062 0066 0069 0081 0089 0097
                         0133 0143 0148 0155 0160 0182 0201 0220 0226 0286 0290
                         0294 0300
TDATA     0001    0016   0088 0159 0164
TERM      0180'   0201   0199
TMAIL     0227'   0311   0316
TMSG      0224'   0307   0028 0041 0061 0068 0083 0091 0099
UNWIND    00FB'   0135   0138
```

A parallel I/O port of the microcomputer must be connected to the line monitor communications channel tap. If the parallel I/O port does not have a +5-V output, then the power for the line monitor must be supplied by some external 5-V supply. The Transmit Data output of the tap must be connected to the bit D_0 input of the parallel port.

The test terminal must be connected to the switched side of the tap. To determine which side is the switched side of the line monitor channel tap, turn on the power to the equipment and hit any key on the test

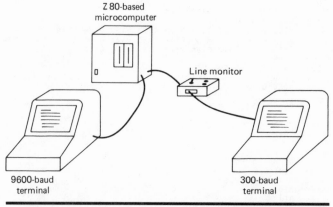

Fig. 4-15   Equipment configuration for Experiment 4-1.

terminal that causes data to be transmitted (e.g. NOT the shift key). Observe that one of the LED indicators on top of the tap will flash. Now flip the line-crossing switch on the tap to the opposite position and hit another key on the terminal. If the same LED flashes as before, the terminal is connected to the unswitched side. If the other colored LED flashes, then the terminal is connected to the switched side.

After ensuring that the terminal is connected to the switched side of the tap, load and execute the software assembled from Listing 4-1. The software console driver address and the parallel I/O port addresses may have to be changed to conform to your system configuration.

Step 2

Via the following prompt, the line monitor software asks you for the number of data bits per character that will be transmitted over the monitored communications link.

Enter bit count per character in hexadecimal notation: 9, A, B, or C.

The response to this prompt must be in hexadecimal notation in the range 9 to C (9 to 12 in base 10). Table 4-5 can be used as a guide in determining the total number of bits used for a given configuration.

Note that if you wish to enter hex A, B, or C, you must use uppercase letters.

The second prompt will be for the number of data bits being observed in the transfer:

Enter data bits per character: 6, 7, or 8.

The response to this must be 6, 7, or 8. This number is the total number of bits per character minus the start, parity, and stop bits.

TABLE 4-5 Number of bits per character for the possible asynchronous character format options

Start bit	Data bits	Parity bit	Stop bits	Total	Hexadecimal notation
1	6	Yes	2	9	9
1	7	No	1	9	9
1	7	Yes	1	10	A
1	7	No	2	10	A
1	7	Yes	2	11	B
1	8	No	1	10	A
1	8	Yes	1	11	B
1	8	No	2	11	B
1	8	Yes	2	12	C

A third prompt will ask whether or not to enable the real-time display:

> Enter 1 to enable real-time display.
> Enter 0 to disable real-time display.

As mentioned before, the real-time display can be used if you have a terminal that will display at 9600 bps or at any baud rate if the transmitted character bursts include no more than about four consecutive characters.

Let us first set up the transfer for (hex) A total bits, 8 data bits with the real-time display on. This means that the test terminal will have to be placed in the 8-data-bit, no-parity, 1-stop-bit mode.

The line monitor program will next display a command menu with four user commands.

```
                    MONITOR COMMANDS
••••••••••••••••••••••••••••••••••••••••••••••••••••
1—RECORD ON
2—PLAYBACK
3—CHANGE CONFIGURATION
CTL C—RETURN TO CDOS

Any key—PAUSE DURING PLAY or RECORD OFF
Any key—RESUME AFTER PAUSE
••••••••••••••••••••••••••••••••••••••••••••••••••••
```

The first of these commands is the RECORD command, which is invoked by typing 1 CR on the console terminal. After putting the line monitor in RECORD mode, hit the A key on the test terminal. If no bit pattern is displayed on the console, flip the line-crossing switch on the line monitor box and hit the A key on the test terminal. One of the switch settings should yield a display identical to the following:

> RECORD is on
> 0100000101

Starting with the leftmost bit, we can analyze the bit pattern displayed. The first bit, the start bit, indicates to the receiving device that the line has just gone from an inactive (idle) state to an active state. As stated in Chapter 2, the start bit is always a logic level 0. The next 8 bits are the data bits. The leftmost data bit is the least significant bit, and the rightmost is the most significant bit. The hexadecimal representation of the data bit pattern is 41. Looking up hex 41 on an ASCII conversion table will show that this figure represents the uppercase letter A. This was the character that you typed on the test terminal. The last, or rightmost, bit on the display is a logic 1. For asynchronous data transfers the last bit transferred is the stop bit, which is always a logic 1.

Step 3

Next hit the B key on the test terminal. The console display should now look like this:

RECORD is on
0100000101 0010000101

The second character is almost like the first. Only the second and third bits from the left have changed. If we strip off the start and stop bits, we have the binary representation of hex 42. Referring to a hexadecimal to ASCII conversion table, you can verify that this is the representation of the character B.

Hit the Z key on the test terminal. The console display should now look like this:

RECORD is on
0100000101 0010000101 0010110101

Again stripping off the start and stop bits, we can verify that the line monitor correctly sensed and displayed the ASCII character Z.

Step 4

Now try transmitting some nonprintable characters, such as CTL A, CTL B, and CTL C. You will see that these characters will result in the hexadecimal representations 01, 02, and 03, respectively. Try some other characters and see what the binary representations look like.

Step 5

Next, clear the line monitor input buffer by typing any character on the line monitor's display terminal (not the test terminal). This will cause the line monitor to exit RECORD mode and display the command menu. Now enter a RECORD command (1) and hit the keys A, B, C, and D on the test terminal. The console display should now look like the following:

RECORD is on
0100000101 0010000101 0110000101 0001000101

Now place the monitor in the command mode again by hitting any key on the monitor display terminal. After the command menu is displayed, put the monitor in the PLAYBACK mode (2). The display should look like the following:

```
RECORD is off
A 41 0100000101    B 42 0010000101
C 43 0110000101    D 44 0001000101
```

As you can see, the display now shows the ASCII character, followed by its associated hexadecimal representation, which is in turn followed by the binary representation, including control bits. The obvious feature of the line monitor PLAYBACK function is that it automatically displays the ASCII and hexadecimal representations. This is a very useful feature in diagnosing troubles on a communications line.

Step 6

Reconfigure the line monitor by invoking the CONFIGURATION CHANGE command (3). Initialize the line monitor for (hex) B total bits and 8 data bits with the real-time display on. Put the monitor in the RECORD mode. Set the test terminal up for 8-data-bit, odd-parity, 1-stop-bit operation. Hit the A, B, C, and D keys on the test terminal. The monitor console display should look like the following:

```
RECORD is on
01000001011   00100001011   01100001001   00010001011
```

Comparing this display with the one in Step 5, we see that an extra bit, the second bit from the right, has been added to each character. This bit is the parity bit. Not counting the stop bit, we see that the parity bit is set anytime that the data bit pattern contains an even number of set bits. This even number plus the parity bit results in an odd number of set information bits. This is an application of the *odd*-parity error-checking convention.

While terminals are usually switch-selectable between odd and even parity, odd parity is the most common convention. In the display, the characters A, B, and D contain an even number of set data bits. Accordingly, the parity bit is set to logic 1 for these characters. Since the code for C already contains an odd number of data bits, the parity bit is at logic 0, i.e., not set.

Step 7

Place the monitor back in the record mode. Change the parity option on the test terminal to even. Hit the A, B, C, and D keys on the test terminal. The monitor display should now look like the following:

```
RECORD is on
01000001001   00100001001   01100001011   00010001001
```

Comparing this display to the one in Step 6, you can see that the only difference is the parity bit, again the second bit from the right. Now the parity bit is set only when the total number of set data bits is odd. The characters A, B, and D, which have even numbers of set data bits, do not have their parity bit set. The character C has its parity bit set because it contains an odd number of set data bits.

The line monitor used for this experiment can be enhanced to perform some of the functions of its more expensive commercial counterparts. You may even wish to convert the polling-based line monitor software to interrupt-driven operation. This will greatly increase the capabilities of the monitor. Taking advantage of the Z80's interrupt capabilities will allow this system to monitor higher communication line rates. The ability to monitor both the Transmit Data and Receive Data lines is within the capability of a Z80-based system using the software for this experiment. These enhancements can turn the system into a very useful diagnostic and development tool for communications projects.

Experiment 4-2: I/O Timing Issues

Purpose

The purpose of this experiment is to investigate timing issues related to performing I/O. In the course of this investigation, you will observe the behavior of several Z80 assembly language programs, ranging from I/O "drivers" that implement no timing functions to drivers that use polling and interrupt-driven control techniques to coordinate data transfers with UARTs, USARTs, SIO circuits, and other similar devices.

Step 1

Load the following program. The program is written in Z80 assembly language mnemonics using an I/O port assignment of hex 01 as the data port. This port should correspond to the console port on the system that you use; consequently, it may be necessary to change the actual port number.

```
PROG1:    LD A,41H      ;move ASCII A to the
LOOP:     OUT (01),A    accumulator
          JR LOOP       ;output to console port
          END PROG1     ;repeat this forever
```

Step 2

Begin execution of this program. You should observe that an endless sequence of the ASCII character A is output to the console terminal. It appears that nothing unusual is taking place. However, you shall observe in the next step that you are seeing only a small fraction of the characters that are actually being transmitted to the console terminal. For example, if your console terminal is running at 300 baud, then you can analyze the situation as follows. Determine the total execution time required to pass through the above program loop once. For a 2.5-MH Z80, we determined that the total execution time is 9.2 μs.

Instruction	T cycles
OUT (01),A	11
JR LOOP	12
Total	23

Thus, once every 9.2 μs, the character A is transmitted to the console terminal. We will assume 1 start bit, 7 data bits, 1 parity bit, and 1 stop bit for a total of 10 bits per character. If the console terminal is set to operate at 300 baud or 300 bps, it will be able to handle 30 characters (at 10 bits per character) each second. The display driver that you have just executed is sending one character to the terminal every 9.2 μs. This constitutes a data rate of about 108,695 characters per second. Thus, approximately 1 character is displayed for every 3623 characters transmitted. Clearly, the CPU is considerably faster than a 300-baud terminal.

Suppose the console terminal is capable of operating at 19,200 bps. Go through the same analysis that was performed above, and determine how many characters are not being displayed during each second of operation.

At 10 bits per character, a 19,200 bps display can handle 1920 characters per second, but the Z80 CPU is sending 108,695 characters per second to the console terminal. Thus, 1 character for every 56 characters transmitted actually gets displayed. Even at 19,200 bps, the console terminal is 56 times slower than the CPU.

The conclusion is that, if we wish to transmit data from a Z80-based microcomputer to a standard terminal, we will have to implement some mechanism to ensure that there is no loss of data. Essentially, we must slow down the CPU's output rate to one with which the terminal can keep pace.

Step 3

Change the value of the ASCII character in the accumulator to hex 00, and execute the program. You should observe that nothing is displayed on your console screen. The ASCII character corresponding to hex 00 is the NUL character, which is a *nonprintable* control code. Similarly, if you load the accumulator with any character whose hexadecimal value is less than or equal to hex 20, no display will result, since all ASCII characters in that range are control (nonprintable) codes, rather than graphic (printable) codes. All other ASCII characters are printable with the exception of DEL, which has a hexadecimal value of 7F.

Load the accumulator with hex 07, and execute the above program. You should observe that a continuous tone is emitted from the terminal. The ASCII character that corresponds to hex 07 is the BEL character, which will ring the terminal bell at 30 times per second if you are operating at 300 bps. This rate of ringing the bell will sound to the human ear like a continuous tone. If you change the baud rate from 300 to 19,200 bps, you will probably not be able to tell the difference in the ringing of the bell, although at the larger baud rate the bell is ringing 1920 times per second, as compared with 30 times per second at the smaller baud rate.

Step 4

Load the following program:

```
PROG2:    LD A,41H        ;load accumulator with ASCII A
LOOP:     OUT (01),A      ;output this character
          INC A           ;increment the accumulator
          JR LOOP         ;repeat this forever
          END PROG2
```

The only difference between PROG1 (from Step 1) and PROG2 is that the hexadecimal value of the character that is being output from the accumulator is being changed each time that an output occurs. In fact, since the hexadecimal value is being incremented by one each time, the natural collating sequence of the ASCII character set should be output to the console terminal.

Step 5

Execute the program of Step 4 above. What did you observe? Did you observe the collating sequence for the ASCII character set? You should not be shocked that the display generated is a hodgepodge of characters and control codes. The reason for this is evident from the calculation performed in Step 2. This calculation showed that we could expect to see (approximately) 1 character out of every 3623 that were transmitted to a 300 bps terminal or 1 character out of every 56 that were transmitted to a 19,200 bps terminal. We say "approximately" here because we must take into account the time necessary to execute in PROG2 the INC A instruction, which was not present in PROG1. You should now be able to perform this calculation precisely. If you are using a terminal with a baud rate set somewhere between 300 and 19,200 baud, you can calculate exactly what percent of the characters transmitted are actually displayed on your console.

Step 6

Load the following two programs:

```
PROG3:    LD A,41H        ;load accumulator with ASCII A
LOOP:     OUT (01),A      ;output to console port
          INC A           ;increment accumulator
          LD BC,0A01H     ;value for the delay loop
          CALL DELAY      ;subroutine to kill time
          JR LOOP         ;repeat this forever

DELAY:    PUSH AF         ;save accumulator ASCII character
STALL:    DEC BC          ;decrement the delay counter
          LD A,B          ;test for BC = 0000
          OR C
          JR NZ, STALL    ;delay some more if not 0000
          POP AF          ;return when BC gets to 0000
          RET
          END PROG3
```

Step 7

PROG 3 outputs one character, then waits for some time period determined by the count placed into the BC register, and subsequently outputs another character. In principle, if we perform the calculations properly, we should be able to determine what value to place in the BC register pair to output data from the CPU at precisely the console terminal's data throughput rate. We will use 300 bps as an example.

In the DELAY loop, the CPU is *busy waiting*. No real work is being done. The CPU is simply waiting, counting out some time until the slower terminal is prepared to accept another character.

To determine what value to load in BC, we must calculate the amount of time that each of the above programs takes to execute. The number of T cycles for each instruction in the above two programs is given in the following table:

	Instruction	T cycles
PROG3:	LD A,41	doesn't matter
LOOP:	OUT (01),A	11
	INC A	4
	LD BC,xxxx	10
	CALL DELAY	17
	JR LOOP	12
	TOTAL	54
DELAY:	PUSH AF	11
STALL:	DEC BC	6
	LD A,B	4
	OR C	4
	JR NZ,STALL	12
	POP AF	10
	RET	10

If you examine programs PROG3 and DELAY, you will see that there are a total of 85 T cycles that are fixed and independent of the number of times that the STALL loop is executed. The STALL loop itself takes 26 T cycles for each execution except for the last time through the loop, when the JR NZ,STALL instruction takes 7 T cycles instead of 12. Thus, we have an equation for the total delay as a function of the number of times that the STALL loop is executed:

Total T cycles = $(26 * N) - 5 + 85$

where N is the number of times that the loop STALL is executed. This is the same as

Total T cycles = $(26 * N) + 80$

If we assume a system clock of 2 MH, then we can express the above equation in terms of microseconds of delay between outputs as a function of the variable N:

Microseconds between outputs = $(13 * N) + 40$

For a 300 bps terminal we would like to have ⅟₃₀ s or 33,333 μs between outputs. Thus we can solve for N as follows:

$33,333 = 13 * N + 40$

or $N = 2561$ (decimal) counts for the BC register.

We went to a lot of trouble to calculate all the times that contributed to this total delay. However, not all of them turn out to be very important. The instructions that are executed only once take so little time *relative* to the total that we can effectively forget about them. Thus, let us make life simpler for ourselves (at least at 300 baud) and forget about the 40 μs. The time period of 40 μs *compared* to 33,333 μs is only slightly more than 0.1%, so let's forget about it. Thus the simplified equation becomes:

Microseconds between outputs = 13 * N

so N = 2564. The decimal number of 2564 translates to a hexadecimal value of 0A04.

Step 8

With BC = 0A04, execute PROG3 given in Step 7. You should observe that the natural collating sequence of the ASCII character set is output to the console terminal, in order:

ABCDEFGHIJKLMNOPQRSTUVWXYZ [\ ] ^   abcdefghijklmnopqrst uvwxyz. . .

Since the CR (Carriage Return) and the LF (Line Feed) are ASCII characters which are output to the terminal, you should notice that they are interpreted by the terminal as control characters and duly executed. Similarly, the other control characters are performed. In particular, you should hear the bell every so often. For many terminals, SUB (hex 1A) or CTL Z is the Clear Screen control character. Possibly other control functions will be performed as the associated ASCII character is output to the console terminal. In our case the following output was repeated over and over again:

!''#$%&'( ) * + ,-. / 0123456789:;< = >?@ABCDEFGHIJKLM NOPQRSTUVWXYZ[\ ] ∧ abcdefghijklmnopqrstuvwxyz{}˜

You should note that LF precedes CR and that BEL precedes LF. This is precisely what you would expect if you were to look at the collating sequence of the ASCII character set.

Step 9

Experiment with the value that is loaded into the BC register pair to determine how much it can be changed from the calculated value of hex 0A04 without disrupting the ASCII collating sequence display. This should be done in two separate steps. First increase the timing constant and see what happens; then decrease the timing constant and see what happens.

Step 10

If your console supports another data rate, change to it and calculate the timing constant in the BC register pair that is required to support the new bit rate. Then determine experimentally if the calculated value of the timing constant is correct.

Step 11
Load the following program:

```
PROG4:    IN A,(01)       ;input from console keyboard
          OUT (01),A      ;output this character to console
          JR PROG4        ;repeat this forever
          END PROG4
```

This program simply loops endlessly, choosing keyboard characters at random to display on the console terminal.

Step 12
Execute the above program. Since the display will continuously feed characters to the console screen, it is best to execute the above program at a relatively low baud rate, such as 300 baud. Type a character on the keyboard. This character should be output to the console display repeatedly. The character will change only after a different character is typed.

After trying a few of the printable ASCII characters, type a few of the nonprintable control characters. The control functions are not associated with the same keys on every keyboard. However there are a few control characters that are reasonably standard. For example, type the control character for the BEL function, which corresponds to hex 07 and is frequently CTL G. (Hold down the key labeled CTL simultaneously with typing the character G.) This will send an endless string of BELs to the console terminal. You will get tired of the noise very quickly, so type another key to get rid of it.

Fill the screen with some characters and then type such cursor control characters as CTL H (BS, Backspace), CTL L (FF, Form Feed), CTL K (VT, Vertical Tabulation), CTL J (LF, Line Feed) and CTL M (CR, Carriage Return).

Step 13
If your terminal allows access to switch settings to control such communications parameters as number of data bits per character, bit or baud rate, parity sense (even or odd or none), and number of stop bits, change these settings one at a time and observe how the output of PROG4 is affected.

Experiment 4-3: I/O with a Programmable Communications Interface

Purpose

In the previous experiments we have looked at I/O that was attempted without any timing control and I/O for which the timing control was supplied (in software) with the use of a software-implemented variable-length delay loop. I/O without timing control was clearly disastrous, and while I/O with timing supplied in a DELAY loop does function properly, the CPU was kept *busy counting* and could not do any productive work. Counting is a trivial function that can be performed easily and very inexpensively in hardware. In this experiment we will examine components that perform this timing function, in addition to performing many other functions.

Our goal is eventually to look at interrupt-driven I/O techniques. In the process we have been looking at alternative I/O methods so that you will appreciate the interrupt-driven technique when you see it. The first step in that direction is to study the components that perform the timing function that was implemented in software in the previous experiment.

Equipment

In this experiment we will look at performing I/O with the timing supplied by a separate I/O component, such as the Z80 SIO, or the Intel 8251 USART, or one of the many programmable serial communications interface circuits that are available.

Step 1

Load the following program:

```
PROG5:    CALL UART       ;UART initialization
          LD C,00H        ;console port
LOOP:     CALL GETC       ;input a character
          CALL PUTC       ;output the character
          JR LOOP         ;repeat forever
                  Subroutines
UART:     LD A,84H        ;300 baud, 2 stop bits
          LD C,00H        ;port for baud rate out
          OUT (C),A       ;output the rate
          LD A,09H        ;reset command word
          LD C,02H        ;command output port
          OUT (C),A       ;output
          RET
PUTC:     PUSH AF         ;save output character
```

```
LOOK:     IN A,(C)         ;look at the transmitter buffer empty flag
          RLA              ;rotate it to carry bit
          JR NC,LOOK       ;repeat until buffer empty
          INC C            ;change to data output port
          POP AF           ;restore character to A
          OUT (C),A        ;output the character
          DEC C            ;restore C to original
          RET              ;return to main
GETC:     IN A,(C)         ;examine data-available bit
          RLA              ;rotate it to carry bit
          RLA
          JR NC,GETC       ;repeat until character available
          INC C            ;change to data input port
          IN A,(C)         ;input the data
          RES 7,A          ;make it ASCII
          DEC C            ;restore C to original
          RET
          END PROG5
```

The program consists of one main routine called PROG5 and three sub-routines, UART, PUTC, and GETC. These routines will each be discussed separately.

PROG5   The main routine initializes the UART and then enters an endless loop in which it reads a character from the console keyboard and echoes back the character just read to the console display.

UART   The UART routine programs a TMS 5501 UART to operate at 300 bps with 2 stop bits. The exact parameters (or control bytes) that must be written for a particular circuit will vary. Each USART, SIO, etc., is programmed slightly differently, but the general process is the same. Programming the circuit simply initializes it for the particular mode of operation that you desire. Once initialization is accomplished, the device takes over and performs the timing and serial to parallel conversion without help from the CPU.

If your system uses a programmable serial interface circuit other than the TMS 5501 programmable I/O controller, then you will have to modify the UART subroutine. Refer to the technical data sheet for your particular device to obtain programming information.

PUTC   This routine is responsible for outputting the character in the accumulator to the console display. The routine shows how the timing information is communicated from the UART (or USART or SIO or similar device) to the CPU. The timing is communicated by way of the Transmitter Buffer Empty flag. The UART simply sets this flag to a logic

1 value whenever the transmission of the previous character has been accomplished at the required bit rate. In other words, the UART serializes the parallel data byte, adds the framing bits and parity bits if desired, and clocks the bits out onto the transmission line at whatever bit rate it is programmed to operate. The Transmitter Buffer Empty flag serves as the synchronization mechanism for the UART to inform the CPU that the previous character has been transmitted and that a subsequent character may be loaded for serialization and clocking out to the transmission line.

The routine PUTC goes into a loop with the CPU constantly asking the UART:

"Is the Transmitter Buffer Empty?"
"Is the Transmitter Buffer Empty?"

Until the UART answers affirmatively, the CPU does nothing except continue to ask this question. Once the question is answered affirmatively, the CPU proceeds to output the character and then returns to the main routine. Thus, while the CPU is *not* performing the timing function directly, it is *busy waiting* on the Transmitter Buffer Empty flag. The CPU could be performing other tasks between questions to the UART, but it has the responsibility of asking the UART for "permission" to transmit the next character.

GETC   This routine is quite similar to PUTC with the exception that it is responsible for input rather than output. Notice that once GETC is entered, control will remain within GETC until a character is detected at the input port by the UART. The question being asked by the CPU is:

"Has a character arrived at the input port?"
"Has a character arrived at the input port?"

This question is repeated until the UART responds with an affirmative answer. The name usually associated with the above question is the Receiver Data Available flag.

Step 2

Begin execution of PROG5. Type a test sentence on the console terminal, such as:

Now is the time for all good men to come to the aid of the party.

This test sentence should be echoed back character by character to the console terminal. Type a few characters and follow them with a CR. You may notice that the character echoed to your terminal upon the receipt of a CR is (strangely enough) a CR and nothing else. In particular, the transmission of a CR does not result in the reception of a CR

LF. To receive CR LF, you must transmit CR LF, unless the software is specifically programmed to echo CR LF in response to transmission of a CR.

One should probably use at least two different test sequences and attempt to alternate the sequences that are used from time to time. You want to make sure that you are observing the results of the current test and not simply the results that were left over from the previous test, when you entered precisely the same test sentences in precisely the same order. Make a typing mistake or do something similar to change the test sequences deliberately.

Experiment 4-4: Interrupt-Driven I/O

Purpose

The purpose of this experiment is to provide an example of interrupt-driven I/O. The experiment will demonstrate the execution of a background task that can be interrupted by an asynchronous request to perform a function that is unrelated to the function being performed by the background task.

Equipment

In order to conduct this experiment you will need access to a Z80-based microcomputer system with a terminal console and a separate I/O device such as a second terminal or a serial character printer. The microcomputer system must support the Z80 mode 2 vectored interrupt structure and be equipped with an interrupt-driven serial I/O port. If you are using a TRS-80 to perform this experiment, you will need to fabricate the circuit that is described in Experiment 4-5. (Experiment 4-5 demonstrates the assembly of a vectored interrupt-driven serial I/O port for the TRS-80. This assembly does not require any hardware modifications to the TRS-80 itself.)

Software

The following listing represents the software required to perform this experiment. The software is further documented in the discussion that follows the listing itself.

```
;
;                    Main program (background task)
;
BEGIN:      CALL INIT
LOOP:       IN A,(00H)          ;get console status
            AND 040H            ;character available?
            JR Z,OUT            ;if not, skip to output
            IN A,(01H)          ;if so, get character
            AND 07FH
```

```
                        CP 03H                  ;was it a CTL C?
                        JR Z,EXIT               ;if so, abort program
                        CP 11H                  ;was it DEVICE ON?
                        JR Z,DEVON              ;if so, branch
                        CP 13H                  ;was it DEVICE OFF?
                        JR Z,DEVOFF             ;if so, branch
                        LD (STORE),A            ;otherwise, store it
            OUT:        IN A,(00H)              ;get console status
                        AND 080H                ;is Transmitter Ready?
                        JR Z,OUT                ;if not, try again
                        LD A,(STORE)            ;get character to output
                        OUT (01H),A             ;output character
                        LD (STORE),A            ;save character
                        JR LOOP                 ;and do it again
            DEVON:      EI                      ;enable interrupts
                        LD A,2AH
                        OUT (0C1H),A            ;output a NUL to start I/O
                        JR LOOP                 ;do it again
            DEVOFF:     DI                      ;disable interrupts
                        JR LOOP                 ;and do it again
            EXIT:       DI                      ;disable interrupts
                        JP 0000H                ;and abort


                             ;Interrupt service routine
            LPOUT:      EX AF,AF'               ;exchange registers
                        LD A,(CHAR)             ;get character for printer
                        INC A                   ;bump it one character
                        AND 07FH                ;mask off top bit
                        OUT (0C1H),A            ;and output it to printer
                        LD (CHAR),A             ;store character
                        EX AF,AF'               ;restore registers
                        EI                      ;reenable interrupts
                        RETI                    ;and return from interrupt


                             Initialization routine
            INIT:       LD A,84H                ;initialize I/O for 300 baud
                        OUT (0C0H),A
                        LD A,09H                ;load device command word
                        OUT (0C2H),A
                        LD A,020H               ;load device interrupt mask
                        OUT (0C3H),A
                        LD A,00H                ;disable unused device interrupts
                        OUT (050H),A
                        LD HL,TABLE             ;load address of vector table
                        LD A,H
                        LD I,A                  ;initialize interrupt register
                        LD IY,LPOUT
                        LD (TABLE),IY           ;initialize vector table
                        IM2                     set Z80 mode 2 vectored interrupts
                        RET
            STORE:      DEFB 2AH                ;console output character
            CHAR:       DEFB 30H                ;printer output character
            TABLE:      EQU 015AH               ;vector table address
                        END
```

Functional Description of Software

The software can be partitioned into three functional units: the *main program*, which executes the background task; the *interrupt service routine;* and the *initialization routine*. The primary function of the *main program* is to output a continuous stream of ASCII characters to the console terminal. The initial character transferred to the console is the asterisk. The character output to the console terminal may be changed by typing any key on the console keyboard with the exception of three special control keys. These special control keys, discussed later, serve to enable and disable the interrupt processing and to terminate the program and return to the operating system.

The *interrupt service routine* repeatedly outputs the ASCII collating sequence to a second terminal or serial printer. The system interrupts must be enabled in the main program before the interrupt routine can be executed. The enabling and disabling of the system interrupts is performed under operator control.

The *initialization routine* is executed once at the beginning of the main program to initialize the interrupt structure with respect to both the Z80 and the interrupt-driven I/O port. Once executed, this routine will not be called again.

Each of these three major functional components of the software is discussed in more detail below.

Initialization   This routine may need to be customized for your particular system. The code shown was written for a Cromemco system using a TU-ART I/O board having a device A base address of hex 50 and a device B base address of hex C0.

The initialization routine performs the following functions:

- Sets the baud rate of the device to be operated in an interrupt-driven mode.
- Enables the Transmitter Buffer Empty interrupts and resets them on the proper port.
- Disables other interrupts from that port.
- Initializes the Z80 interrupt register with the high-order address byte of the interrupt vector table.
- Initializes the vector table with the address of the interrupt service routine.
- Sets the Z80 to operate in the vectored interrupt mode (mode 2).

Thus, the initialization routine sets up the interrupt mechanism for both the Z80 and the I/O port so that an interrupt will be generated each time that the Transmitter Buffer Empty flag in the I/O device is active. When this occurs, the I/O device will gate an interrupt vector onto the system bus. In the case shown here an interrupt vector of hexadecimal value 5A will be gated onto the system bus in response to an Interrupt Acknowledge signal from the Z80.

The Z80 interrupt register (I register) is loaded with the high-order byte of the address of the interrupt vector table. We have decided to put this table at memory location hex 100. It could, however, be located on any 256-byte boundary in unused read/write memory. Remember that the low-order byte of the interrupt table address is determined by the vector supplied by the I/O device. In our implementation of this experiment, the Cromemco TU-ART card has been initialized to supply the vector hex 5A. The interrupt vector table must be loaded with the proper interrupt service routine address. In this case, the address of the service routine is determined by the label LPOUT, so this label value is moved to the address TABLE.

Main Routine (Background Task) The background task outputs a continuous stream of characters to the system console. The task polls the console input. If a character is available at the console port, the character will be checked to see if it is a special control character. If it is not, then it is stored in the memory location labeled *STORE*. The task then polls the console output port until it is ready to accept a character. At this time the character previously loaded at the memory location STORE will be transmitted to the console. As long as no new character is received from the console keyboard, the character stored in this memory location will continue to be transmitted.

If the character found at the input port is a DC1 (hex 11), then control is transferred to a short routine which will enable the Z80 interrupts and output a NUL character to the I/O device. This NUL character (it could be any character) is needed to start the I/O device interrupt process. Once this is done, control is returned to the main program. At this point the system is ready to begin interrupt-driven I/O with the next I/O port interrupt request.

If the console input character is a DC3 (hex 13), then control is passed to the section of code labelled DEVOFF. This code will disable the interrupts and return control to the main program. At this point the interrupt-driven I/O will stop.

If the console input character is CTL C, then control is passed to the section of code labeled EXIT, where interrupts are disabled and control is passed to the operating system.

Interrupt Service Routine The interrupt service routine is invoked whenever an interrupt request from the serial I/O port is received. If the console I/O port is transferring at 19,200 baud and the interrupt port is transferring at 300 baud, then this port will request service after approximately 64 characters have been output to the console. Once an active interrupt signal is detected and accepted by the Z80 CPU, maskable interrupts are automatically disabled until they are reenabled under software control. The current state of the CPU registers is saved by the interrupt service routine so that the background task may resume execution in precisely the same state that it was in when the asynchronous interrupt occurred. This is accomplished in this example by exchanging the primary AF registers with the alternate AF registers. There is no need to save the state of the other registers since the interrupt service routine does not alter them. This could also be achieved by storing the contents of the registers on the system stack and then restoring them at the completion of the interrupt service routine. The stack approach is necessary when using CPUs other than the Z80 that do not have the alternate register set. The stack approach is also necessary when it is desired to permit *interrupt nesting*. Interrupt nesting occurs when interrupts can be reenabled within an interrupt service routine so that interrupting an interrupt service routine is possible.

Once the state of the CPU has been saved, the next function of the routine is to retrieve the character stored in the memory location labeled CHAR. The content of this location is then incremented by 1, the most significant bit is masked off, and this new bit string is then transferred out the serial I/O port. Notice that the Transmit Ready bit is not sampled in this routine. This is because the Transmit Ready bit caused the interrupt that invoked the interrupt service routine. Therefore, this section of code is only executed when the serial port transmitter is ready to accept a character. The act of incrementing and masking the most significant bit of the character stored at the location CHAR has the effect of sequencing through the entire 128 characters of the ASCII code set. This then automatically generates the collating sequence of the code set.

After the character to be printed has been loaded into the serial port data register, the interrupt service routine restores the original contents of the AF registers. Again this must be done so that any process in the main program that was interrupted by the serial device service request can continue once control is returned to the main program. Before control is returned to the main program, however, the interrupts must be reenabled so that the next interrupt cycle can receive service. Once this is done, the return from interrupt is executed and the main program resumes.

If this experiment is to be conducted on a system other than the type described above, some modifications to the software may be necessary. In particular, you may have to change the addresses of the console control and data registers. The address of the serial I/O port and the command and control register address assignments will probably also be different for your system.

Step 1

Connect a 300-baud terminal or character printer to the interrupt-driven serial I/O port.

Step 2

Load and execute the software for this experiment. The main program will immediately begin to output asterisk characters to the console. There should be no response on the printer at this time. Hit the S key on the console keyboard. The console should now begin to display S characters on the screen. After the screen is filled with S characters, hit the period key. The screen will now fill with period characters. Next hit the space key. The console screen should now begin to clear as Space characters begin to fill the screen. This clearing effect should be the same for any nonprintable key that you enter on the console keyboard. Now hit the zero key.

Step 3

After the screen has filled with zeros, enter CTL Q. The printer should immediately begin to print the ASCII collating sequence. This is because the CTL Q is the DC1 character, which causes the main program to enable the serial port interrupts. Notice that there is no appreciable decrease in the rate at which the characters on the console screen are being displayed. This is because it only takes a few microseconds out of every 33 ms to output a character to the serial I/O port. Now hit the L key on the console. Observe that the console screen begins to fill with L characters but that the collating sequence being sent to the printer is unaffected. Hit several of the other keys on the console, and notice that the interrupt service routine and the main program appear unaffected by one another. In reality, the main program is slowed down when the interrupts are enabled, but this is not perceivable to the system operator. If several interrupt service routines were active, the degradation of performance would become more noticeable.

Step 4

After a few of the ASCII collating sequences have been printed, type CTL S on the console. The printer should immediately stop. This is because the CTL S is the DC3 character, which will disable the interrupts and stop the flow of data to the serial I/O port. Again notice that no difference is noticeable in the throughput rate of the main program. Try entering characters and see that the screen fills with the different characters being entered.

Step 5

Type CTL C on the console. Control should be passed to the operating system.

This experiment forms the basis for an interrupt-driven printer-spooler. By adding a buffer and some buffer control to the experiment software, the reader can implement a simple line printer-spooler, which enables the printing of a buffer to occur concurrently with the execution of a background task. This interrupt-driven I/O concept is very important in high-performance applications that rely on direct memory access (DMA) for the data transfers. Although interrupt-driven software is not frequently used for low-end hobby and personal microcomputer systems, it is very common for performance-oriented and commercial microcomputer systems. Minicomputers and large mainframe systems use the interrupt-driven I/O technique almost exclusively.

Experiment 4-5: Vectored Interrupts and Interrupt-Driven I/O on the TRS-80 Model I

Purpose

The purpose of this experiment is to implement an inexpensive vectored interrupt-driven serial I/O port on the TRS-80 model I. After fabricating the experiment hardware, we will discuss how the interrupt-driven software from Experiment 4-4 can be modified to work with this new hardware on the TRS-80. This experiment requires *no hardware modifications* to the TRS-80 itself.

Equipment

The following list of equipment and components will be needed for this experiment:

1	TRS-80 model I with Radio Shack editor/assembler
1	Intel 8251 programmable communications interface integrated circuit
1	Motorola MC14411 bit rate generator
1	7430 Eight-Input AND gate
1	7404 hexadecimal inverter
2	74367 tri-state hexadecimal buffers
1	1488 RS-232 line driver
1	1489 RS-232 line receiver
1	Seven-pole DIP switch
8	0.1-μF capacitors
7	10-kΩ 0.25-W Resistors
1	15-MΩ 0.25-W Resistor
1	1.8432-MHz Crystal
1	24-pin wire-wrap socket
1	28-pin wire-wrap socket
6	14-pin wire-wrap sockets
2	16-pin wire-wrap sockets
1	40-pin edge connector (Radio Shack no. 276-1558)

3 ft	40-pin conductor ribbon cable
1	40-pin female ribbon cable connector (Jade no. CNF-62400)
1	40-pin wire-wrap male header (Jamesco no. 923865-R)
1	4 x 6 in project board
1	4¼ x 6¼ in project box
1	spool no. 30 awg wire-wrap wire

Circuit Description

The circuit shown in Figure 4-16 will be used to implement a vectored interrupt-driven serial I/O port on a TRS-80 model I. When completed, the circuit will interface to the TRS-80 via the expansion interface connector on the back of the model I keyboard. This experiment will require no hardware modifications to the TRS-80 itself. This new serial I/O port can be used to interface the TRS-80 to RS-232 equipment like modems, terminals, or serial character printers. When used with a minimum of communications software, this new serial I/O port will significantly enhance the function of a basic TRS-80 system. In addition, this new I/O port can be used to add vectored interrupts to systems that already have the Radio Shack expansion interface and the Radio Shack RS-232 port installed.

The major component of the serial I/O port is the Intel 8251 programmable communications interface. This is an improvement on the standard UART and includes the capability to handle synchronous, as well as asynchronous, communications protocols. The 8251 circuit was originally developed by Intel Corporation, but it is now available also from National Semiconductor and others. This chip performs some of the same functions as the Zilog SIO circuit discussed in Chapter 2. The main function of the 8251 with which we are concerned is parallel to serial conversion.

Looking at the schematic in Figure 4-16, we see that the 8251 data lines connect directly to the TRS-80 data lines. This is the path for all data and communication between the 8251 and the TRS-80. The TRS-80 OUT and IN lines connect to the 8251 Write and Read lines, respectively. This provides control for the direction of data flow. The TRS-80 Reset line is inverted and connected to the 8251 Reset line. Address line A_0 is used to select between the data and command registers of the 8251. Address lines A_1 to A_6 are connected to six inputs of the 7430 eight-input NAND gate. Address line A_7 is first inverted and then connected to the 7430. This leaves one unused input to the 7430, which is tied high to 5 V. This configuration results in an active low output of the 7430 whenever address lines A_1 to A_6 are high and A_7 is low. Therefore the output of the 7430 will enable the active low chip-select input of the 8251 whenever a low byte address of hex 7E or 7F is present on the TRS-80

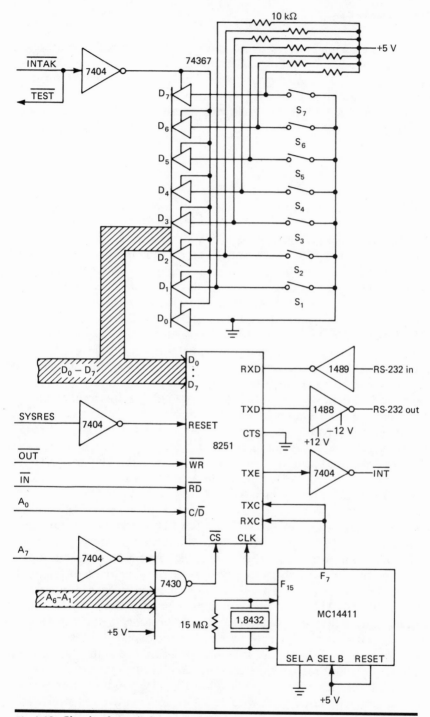

Fig. 4-16 Circuit schematic for Experiment 4-5.

address lines. This address range can be changed to suit the needs of the reader simply by adding inverters between the address lines and the appropriate 7430 gate inputs. Note that I/O address hex FF must be reserved for the use of the TRS-80 system.

Summarizing, we have the following address map for the new serial I/O port:

7EH OUT	Data OUT
7EH IN	Data IN
7FH OUT	Command OUT
7FH IN	Status IN

The only other connection to the TRS-80 from the 8251 is the interrupt request input. The 8251 can supply an active high input on Transmitter Buffer Empty, Transmitter Ready, and Receiver Ready. For the purpose of executing this experiment, we have tied the inverted output of the Transmitter Ready Signal to the active low interrupt request input. The reader may wish to connect the Receiver Ready input to the interrupt request after completing this experiment.

There are three other support chips that interface to the 8251. These are the MC14411 clock generator, the 1488 line driver, and the 1489 line receiver. The MC14411 can be configured to provide many of the standard line frequency clock rates commonly used. For this application we have selected a transmit and receive clock rate of 19,200 bps. The 8251 can internally divide this down to 1200 and 300 bps. The MC14411 needs only a simple crystal circuit and power to provide a very stable clock frequency. The crystal circuit consists of the 1.8432-MHz crystal in parallel with the 15-MΩ resistor. The MC14411 also provides the 8251 with a 900-kHz internal clock.

The 1488 and 1489 circuits provide for level conversion between the communication line RS-232 levels and the TTL levels used by the 8251. Notice that the 1488 line driver will need an external source of ± 12 V.

The most important aspect of this experiment is the capability provided by the circuit in the upper portion of the schematic in Figure 4-16. This is the circuit that allows us to implement vectored interrupts on the TRS-80. The main portion of the circuit is the tri-state noninverting buffers that are used to gate the vector address onto the TRS-80 bus. The input to the tri-state buffers is controlled by the settings on switches S_1 to S_7. When these switches are open, the buffers are pulled high by the 10-kΩ resistors attached to each input. When the switches are closed, the buffer inputs are held low by the grounds on the opposite sides of the switches. These switches then allow the user to specify any 8-bit interrupt vector by setting switches S_1 to S_7. Notice that bit 0 is always tied low. This is to ensure that the vector is always on an even address boundary to be consistent with standard Z80 interrupt conventions.

The final function of the vector gating circuit is to handle the Interrupt Acknowledge signal. This signal comes from the TRS-80 and is active low when the 8251 has requested service and the Z80 processor has completed the current instruction cycle and is ready to accept the interrupt vector. The Interrupt Acknowledge signal is inverted and used to enable the tri-state buffers that supply the vector to the system bus. At the same time, the active low Interrupt Acknowledge signal is sent back to the TRS-80 on the active low Test line. This Test signal will put the internal TRS-80 tri-state bus buffers in the input mode so that the vector can be received by the CPU. In addition, the Test signal tri-states the address bus and makes a bus request to the Z80 CPU. These extra functions have no effect on the vector input cycle and can be ignored for this experiment.

Now that you have seen how the TRS-80 interrupt-driven serial I/O port is designed to work, you are ready to begin to fabricate the hardware.

Step 1

You should begin this experiment by first developing a component layout. This is best accomplished by temporarily placing the wire-wrap sockets on the project board and arranging them in the best pattern to fit the available space. In this experiment the size of the project board and the small number of integrated circuits (ICs) make this a relatively easy task. In fact, the layout of the components and the size of the project board are not critical for this experiment. You should only be concerned that the finished board fits into the selected project box. Once the components have been laid out, the wire-wrap sockets should be epoxied into place. There must be one socket for each IC, one for the seven-pole DIP switch, and one for the seven 10-kΩ resistors. Experience dictates that it is highly desirable to use high-quality IC sockets. This eliminates hours of troubleshooting the erratic problems that arise when cheaply manufactured IC sockets are used. You are also cautioned against using fast-drying glues. These glues tend to flow into the pin cavities and prevent good electrical contact between the IC pins and the socket contacts. It is advisable to use a *small amount* of a two-part epoxy for attaching the sockets to the project board.

Step 2

Once the IC sockets have been mounted, the 40-pin male header should be mounted on the project board. Install this connector close to one of the edges of the card. Epoxy the header into place. The 40-pin card edge connector and the ribbon cable connector should now be crimped onto the 3-ft length of ribbon cable. This process is best done with a mechanical crimping device. Since these devices are not readily available, a small woodworking vise has been tried and used quite successfully by the authors. In any case this is a nontrivial task, and the help of someone experienced in installing this type of connector should be recruited. It may even be advisable to purchase cables that have already been terminated with the desired connectors, even though such cables are fairly expensive.

Step 3

After installing the IC sockets and preparing the cable, you are ready to begin wire-wrapping the circuit. *Please note* that the pin assignments on the card edge of the TRS-80 keyboard are different from the normal assignments. If you look at the card edge of the keyboard from the rear, the odd-numbered pins will be on the top and the even-numbered pins on the bottom. Normally the reverse is true. This means that the ribbon cable tracer will be on pin 2, not pin 1. This can be corrected at the header on the project board by again reversing the normal pin assignments. It is suggested that the pin assignments be checked out with an ohmmeter from board to board when the cable is installed. After marking the correct pin assignments for the header on the project board begin wiring the circuit shown in Figure 4-16.

Step 4

The resistors and the crystal circuit can now be installed. The 10-kΩ resistors will fit nicely into the spare 14-pin DIP socket. The crystal and the 15-MΩ resistor will have to be soldered to the proper pins on the 24-pin socket. It may also be helpful to terminate the line driver and receiver with a female 25-pin D connector so that an RS-232 cable can be connected directly to the I/O port circuit.

Step 5

Before installing the ICs, you should connect the power supplies to the circuit and test the wire-wrap sockets for power on the approprate pins. You will need an external power source for +5 and ±12 V for this circuit. Once you have ensured that power is properly applied, remove the power and install the ICs. A final visual check should be made before reapplying power to the circuit.

Step 6

After the serial I/O port hardware has been checked, you can rewrite the software for Experiment 4-4 so that it will execute on the TRS-80.

Some points to remember in making the modifications are outlined below:

- The console I/O segments must be changed to call the TRS-80 system keyboard input and video display routines resident in the system's read-only memory (ROM).
- The control characters must be changed to some other character because the TRS-80 has no control key. One approach might be to use shift Q to jump to the DEVON routine, shift S to jump to the DEVOFF routine, and shift C to branch to the exit segment. TRS-80 BASIC will interpret these shifted characters as lowercase letters.
- The EXIT segment should branch to hex 1A19, as the system reentry point.
- The interrupt service routine, LPOUT, must be altered to output data to I/O address hex 7F instead of hex C1.
- The INIT routine must output hex 4F to the command out address hex 7F, followed by hex 15 to the same address. This initializes the serial port for 1-stop-bit, no-parity, 8-bit characters, 300 baud, and enables the transmitter and receiver.
- The vector address is set by the hardware, so it is up to you to set it for the desired address.
- The I register and the vector address table will need to be initialized by the INIT routine. You must keep in mind that the user read/write memory in the TRS-80 is above address hex 42E9. Thus, all of the software and the vector table must be located above this address.

Although these changes may look extensive, you will find them to be relatively easy to make. Once you have gained some experience in vectored interrupt-driven I/O, you will see that this technique can greatly increase the throughput of systems like the TRS-80.

5 *Modems*

Modems provide the means for economically extending the reach of modern computer systems. By providing an inexpensive means for communications, modems can move a multimillion-dollar computer complex and its associated computing resources from a remote facility to an office, a laboratory, a retail store, or a private home. Indeed, for an investment of under $1000 it is possible to purchase a modem and terminal and to tap the resources of remote computing facilities by making a telephone call.

MODEM APPLICATIONS

In this chapter, we investigate modems and their crucial role in communications via the public-switched telephone network. In particular, we will address what functions modems perform, why these functions are necessary for communications over the telephone network, what types of modems are commercially available, how modems work, and finally, what the characteristics are of the public-switched telephone network with respect to data communications capacity, interfacing, and transmission errors. Before we begin to discuss these topics, however, let us first look at several modem applications that are having a profound impact upon our society as we enter the information age.

Information Networks

The capability of voice communications is established in virtually every U.S. home and is becoming increasingly established in the homes of every industrialized country worldwide. The microelectronics revolution is extending this capability to data communications, as well. Merely a decade ago, computers and computer terminals in the home were economically and practically infeasible. This is no longer the case.

To capitalize on this incredible new market, many large companies and groups of smaller companies are developing products and services for the home, including computer games and entertainment packages, at-home banking and shopping, educational packages, on-line telephone directories, travel information, classified advertisements, and movie and theatre reviews. The companies that are marketing these services range from the American Telephone and Telegraph Company to a toy manufacturer and an electronics company that are the major partners in a joint venture.

In the United States, several information networks are currently available to home users. These networks are often referred to as *videotex* or *viewdata* networks. The major characteristic of these networks is low-cost, two-way communication that is user-oriented (no computer expertise required), with simple access via inexpensive terminals, modems, and the public-switched telephone network. The earliest such network, called Prestel, was developed and marketed in the United Kingdom by the British Post Office. The U.S. viewdata networks include the following:

- *The Source:* A low-cost time-sharing service that is accessible nationwide via the value-added networks Telenet and Tymnet
- *CompuServe Information Service* or *MicroNet:* A spin-off of a time-sharing company's commercial services that uses the company's existing private network (leased from the telephone company)
- *Viewtron:* A joint project of AT&T and Knight-Ridder Newspapers
- *Green Thumb:* A network that provides weather and crop information to farmers, developed by the U.S. Department of Agriculture

Our main purpose in mentioning these services is to emphasize that their use of telephone communications requires modems.

Computer Networks

Today computer networks are being developed that allow many computers to communicate with one another for the purpose of sharing data, sharing workloads, sharing resources, and controlling external functions. Banks use these networks to allow bank tellers to update or check transactions instantly with a central computer. Supermarkets use computer networks to keep track of price changes and to make this information available to automated checkout counters. Stock brokerage firms use computer networks to initiate and track their transactions on the various exchanges. Computer time-sharing services connect computers together and sell their resources to customers who do not have the need or the desire to establish and maintain their own computer centers. There are hundreds of different computer network applications, yet they all share one common need: the need for the communication channels that exist between the network's nodes and the modems necessary to implement these channels.

Telecommuting

Modems are the fundamental component in a relatively recent phenomenon, commuting to the office by way of terminal and modem, otherwise known as *telecommuting*. Credit for coining the word *telecommuting* is usually given to John Nilles, now the Director of Interdisciplinary Studies at the University of Southern California, who first used the word in the early 1970s. Telecommuting is currently being experimented with on a serious basis. John Nilles' current projections are that by the mid-1990s, over 10 million workers will be telecommuting. An obvious advantage of telecommuting is the significant saving of time and energy normally spent in commuting to the conventional physical office. Also, via telecommuting, employers have access to a labor pool that might not otherwise be available.

Remote Diagnosis

Modern minicomputer systems are being delivered today with built-in modems that are used to connect the system to a diagnostic computer often located hundreds of miles away. When a problem occurs or when preventive maintenance is necessary, the computer operator simply dials the phone number of the remote diagnostic computer, establishes a connection, and sits back while the local system is thoroughly tested. If problems are found, the diagnostic computer makes a report to the local field service office, again via a modem and communication line. The local field service office then dispatches a repair technician, who will bring a replacement part if one is needed, having been informed of this by the diagnostic computer. This type of modem application greatly reduces the cost of maintenance of large minicomputer installations.

It is conceivable that in the near future complex systems will include a built-in diagnostic modem. Someday local automobile dealers may have small computer systems connected via a telephone line and modem to Detroit—or perhaps Tokyo. Cars may then be tested by the factory and detailed maintenance procedures sent to dealers through the modem. In fact, this feature may even be extended to the homes of car owners, where they will plug a modem cable into their cars and then will make a phone connection to their local dealers. The cars will be tested and, if any problems exist, the dealers' computers will make an appointment for the car, issue an order for repair parts, and print a repair procedure for the mechanic. When the car is driven in for the appointment, the parts and the mechanic will be ready. All of this will be made possible because of the modem's ability to allow the public telephone network to be used for data communications, in addition to the normal voice traffic.

The feature of most modems that makes them so important for telecommunications is the ability to connect computers together over the public-switched telephone network. The modem does this by converting digital data signals to audio signals that can be transmitted over the public telephone network. At the receiver end, another modem converts

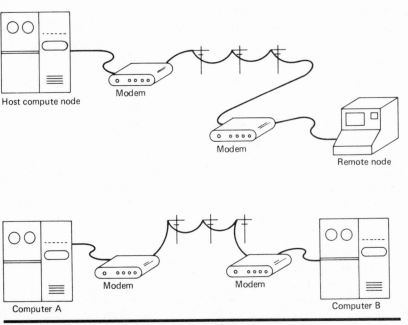

Host compute node

Modem

Remote node

Modem

Computer A

Modem

Modem

Computer B

Fig. 5-1 Modems in teleprocessing configurations.

the audio signals back into digital data signals. This process is called *mod*ulation and *dem*odulation, from which the word *modem* is derived. While many modems connect computers together over private and local lines, by far the vast majority connect systems together over the public telephone network. Figure 5-1 shows two typical teleprocessing configurations that use modems: a terminal-to-computer communications channel and a computer-to-computer communications channel. Utilization of the existing telephone network is quite often the only practical systems design alternative. The cost of running lines, the legal problems involved with crossing private and public properties, and the time required to install lines are insurmountable problems that rule out other terrestrially based alternatives. Even though the public telephone network has some serious limitations in terms of data communications, it exists and it works. Therefore its advantages far outweigh its disadvantages.

OBJECTIVES

At the end of this chapter, you will be able to do the following:

- Explain the function of a modem as a component in a communications system.
- Explain why modems are needed to use the public-switched telephone network.
- Identify and define several categories and types of modems on the basis of their transmission speed and modulation technique.
- Describe a Bell 103 modem and explain how it performs the functions of modulation and demodulation to transmit data to and receive data from the telephone network.
- Characterize the public-switched telephone network in terms of its data communications capacity, interfacing properties, and bit error rates.

The above objectives for this chapter are designed to provide an introduction to modems. Several references are given in the text that address the subject of modems in much greater depth.

MODEM FUNCTIONS

A modem is a device that modulates and demodulates an electrical signal that is used to transfer data between two computing devices, or "peripherals." *Modulation* is defined as the process of modifying some characteristic of an analog signal (called a *carrier*) so that it varies in step with some other analog signal (called a *modulating wave* or *signal*). Figure 5-2 illustrates the function of a modem in a typical communications configuration. In order to transfer data from a computer system to a remote terminal, the data are sent out of a serial I/O port from the computer system. If the data are transported over the public-switched telephone network, then the serial I/O port is connected, via an RS-232-C cable, to a modem. The modem converts the computer's digital signal to a modulated analog signal, suitable for transmission over the phone line to a remote site. At the remote site another modem demodulates the analog signal to extract the digital information. The remote modem then passes this information to a locally connected terminal.

For most applications the functions of a modem are transparent to the user of a computer system. In the configuration shown in Figure 5-2, a user sitting in front of the remote terminal will have the same system interaction as another user located in the central computer room. In many cases the computer software that handles terminal communications is no different for the terminals in the main computer room than for the terminals located hundreds of miles away at the remote site. One exception to this situation is a teleprocessing port that uses a multifunction, autodial, or multiline modem that has programmable functions and control registers accessible to the host system. Autodial

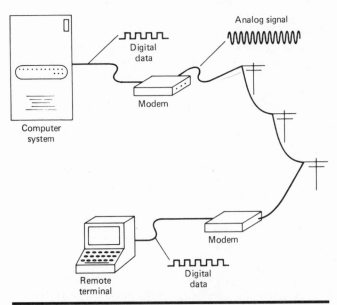

Fig. 5-2 *Modem functions in a communications network.*

modems and autoanswer modems are designed for use in systems for which unattended phone line operation is desired. In these cases the modems can provide the capability to originate and/or answer calls automatically by providing the dialing pulses or tones and by detecting and answering incoming calls. Multiline and multiplex modems provide the host system with the capability to service several incoming lines with only one data channel. Both of these categories of modems will be discussed later in this chapter.

THE NEED FOR MODEMS

The physical characteristics of the lines used by the public-switched telephone network make them unusable for the direct transmission of digital signals. The public telephone service was never intended for data transmission. The telephone lines use audio amplifiers, analog filters, echo suppressors, and other electronic devices that prohibit the use of digital signals. Even the dedicated leased lines available from most common carriers are not suitable for digital signals. State-of-the-art technology dictates that if the public telephone network is to be used for data transmission, then the signals must be converted to the audio frequency spectrum that the network was designed to handle. This frequency spectrum covers the range of 300 to 3300 Hz (or cycles per second). Using inexpensive and simple modems limits the effective transfer rate to around 1200 bps. Data transmission rates above 1200 bps, on the public-switched network, require more complex modems that have an upper limit of about 9600 bps.

CLASSES AND TYPES OF MODEMS

There are many different ways to classify modems. In this section we will investigate two: *speed* and *modulation technique*. When discussing modem speed, there is commonly confusion over the concepts of *bits per second* (bps) and *baud*. *Baud* is defined as the unit of the signaling rate that represents the number of discrete conditions or signal events transmitted per second. *Bits per second* (bps) is defined as the number of binary digits actually transferred per second. The only time when there is a difference between a modem's baud and bit rate capacity is the case in which the modem uses some of the more complex techniques for superimposing digital information onto analog signals. In this case we will see how a baud information element can carry two, three, and even four binary information bits.

Modem Categories

In addition to classifying modems according to speed, we will subclassify them further according to the modulation technique they use. There are other categories that can be used to classify modems, such as the method of connection to the communication lines, or whether they allow full-duplex, half-duplex or simplex communications. However, we shall discuss these features under the two levels of categories previously defined. Figure 5-3 illustrates the hierarchy of modem classes that we will be using.

There are three major categories of modems based upon speed:

- *Low-speed modems*, which transfer data at speeds of 600 bps or less
- *Medium-speed modems*, which operate between 1200 and 9600 bps
- *High-speed modems*, which operate at speeds in excess of 9600 bps

While the majority of modems on the market today fall clearly into one of these categories, there are some modems that are capable of operating in two or all three of these speed ranges. Modems with these capabilities are expensive and complex devices that are not usually present in microcomputer configurations. Therefore, they will not be discussed here.

The second major modem classification that we will consider is modulation technique. The following are the three most common techniques:

- Frequency shift keying (FSK)
- Phase shift keying (PSK)
- Phase amplitude modulation (PAM)

As in the speed categories, most modems fit into one modulation category. However, there are some modems that use a combination of these techniques to superimpose information on an analog signal. There are also other methods of modulating digital signals, but those methods are rarely used in microcomputer systems and therefore will not be discussed. The three modulation techniques listed above are described later in this chapter.

Low-Speed Modems

Let us begin our study of how a modem works by first selecting one type of modem to study, the low-speed modem. After investigating the modulation technique associated with low-speed modem operation, we will then look at the variations and different approaches that other modems use.

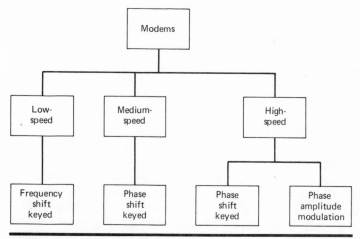

Fig. 5-3 One way of classifying modems.

The Bell 103 Modem

One of the most common modems currently in use is the Bell 103–compatible low-speed modem. Bell 103–compatible modems are among the least expensive and most widely available of all modem types. Hence, 103-compatible modems can be found in almost any large computer installation and have become very popular for microcomputer configurations. Good-quality originate and answer 103-type modems can be purchased for less than $200. While almost every modem manufacturer builds what they refer to as a "103-compatible" modem, this usually means that the modem transmits and receives on the same frequencies as the Bell 103. However, it is important to note that the original Bell 103 modem implemented several features that may not be available on cheaper units, even though they claim 103 compatibility. Some of the features included by the 103 are:

- The ability to operate in originate, answer, or CY lead-controlled mode. (CY lead-controlled operation means the mode can be changed between originate and answer dynamically under software control via the RS-232 unassigned pin 11.)
- The ability to make the incoming phone line appear busy dynamically under software control.
- The ability to disconnect when the incoming carrier is lost.
- The ability to generate the specific tones and/or pulses required by the public-switched telephone network to dial and access a remote station. (This feature is called the *autodial option.*)
- The ability to generate specific signal output levels and meet certain signal-to-noise ratios and operational timing characteristics.

In addition to these features the 103 can be configured to provide many combinations of options that involve the standard RS-232-C modem control signals. The 103 also provides several status indicators. These additional features all cost extra. For small configurations, especially microcomputer systems, less expensive units that do not implement all of the Bell 103 options are quite adequate.

Modulation Techniques

The next paragraphs discuss how the Bell 103 modem works: the frequency assignments that it uses for transmitting and receiving data, the modulation technique that it uses, and the conceptual design of its transmitter (modulator circuit) and receiver (demodulator circuit). In this book, we do not assume that the reader has a background in analog electronics. For this reason, we shall limit our discussion of modem modulation and demodulation techniques to functional descriptions, rather than give detailed specifications. For the interested reader with some background in analog theory, we suggest the following two references for further detailed study of modem circuits:

> John E. McNamara,*Technical Aspects of Data Communications*, Digital Press, Bedford, Mass., 1977
>
> Vess V. Vilips, *Data Modem Selection and Evaluation Guide*, Artech House, Dedham, Mass., 1972

Frequency Assignments (Originate and Answer Modems) As mentioned earlier, the function of modems is to convert digital signals into audio signals suitable for transmission over phone lines. The digital transmission of data relies on the presence of a signal or voltage for representing logic level 1, with another different signal or voltage representing logic level 0. Low-speed modems, however, use two discrete audio signals, one to represent a logic 1 and another to represent a logic 0. This means that four frequencies are needed to communicate on a noninterfering basis in two directions at once (full duplex)—two for the transmitter end (one for logic 1 and one for logic 0) and two for the receiver end. The physical characteristics of analog circuits dictate that these frequencies not be integral multiples of each other. In the case of the 103-compatible modem, in the originate mode, the frequency assignments are:

1070 Hz   Transmit space (logic 0)

1270 Hz   Transmit mark (logic 1)

2025 Hz   Receive space (logic 0)

2225 Hz   Receive mark (logic 1)

These assignments are illustrated in Figure 5-4.

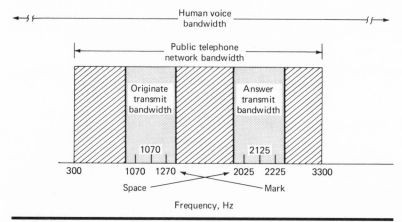

Fig. 5-4 Bandwidth and frequency assignments of low-speed modems on the public-switched network.

These frequency assignments are relative to the modem initiating the communication link, the *originate modem*. The transmit frequency assignments of the modem on the other end of the link, called the *answer modem*, are the same as the originate receive frequencies, and the answer receive frequency assignments are the same as the originate transmit frequencies. This frequency assignment for superimposing digital data on the analog signal is called *frequency shift keying*.

Frequency Shift Keying (FSK)   Figure 5-5 illustrates how FSK is used to represent digital data. The diagram shows the various ways to represent the ASCII character S (hex 53). The top line shows the seven time states used in serial transmission to represent each ASCII character, using the 7-bit ASCII standard code set. (This code set is discussed in detail in Chapter 4.) The second line shows the binary representation of the character S with the least significant bit to the left. The next line is the TTL (transistor-transistor logic) representation of the character. Then comes the RS-232-C representation. The bottom two lines show the FSK-modulated signals that pass between modems. The period of the sine waves of the FSK signals are not drawn to scale. The difference in period of the sine waves is greatly exaggerated to illustrate the different frequencies. On the originate modem Transmit signal, the first two bit times are modulated at a frequency of 1270 Hz, which represents two logic 1 bits. The third and fourth bit times are modulated at a frequency of 1070 Hz to represent two logic 0 bits. The fifth and seventh bit times are also modulated at 1270 Hz, and the sixth bit time is modulated at 1070 Hz. On the originate modem Receive signal, the first two bit times, as well as the fifth and seventh bit times, are modulated

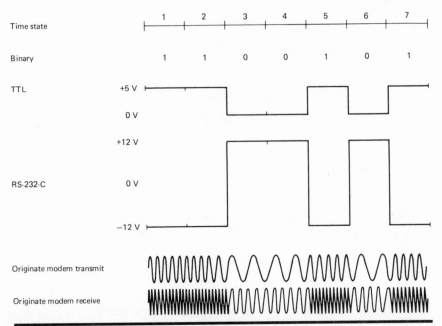

Fig. 5-5 Methods (including FSK) of representing the letter S.

at a frequency of 2225 Hz. The third, fourth, and sixth bit times of the Receive signal are modulated at 2025 Hz.

Modem Transmitters Now that we see what the FSK signals look like on the communication line, let us take a look at how the modem produces these signals. Figure 5-6 is a simplified illustration of how a modem might perform this function. The two oscillators represent the major elements of the transmitter side of a modem. Let us assume that the top oscillator is turned on by a voltage in excess of −5 V and the bottom oscillator is turned on by a voltage in excess of +5 V. If an incoming RS-232-C Transmit Data signal is connected to the input of the oscillators, this signal will turn the oscillators on and off. When the RS-232-C signal is at −12 V, the 1270-Hz oscillator is turned on. Recall that − 12 V is the RS-232-C representation of logic level 1, and an analog signal of1270 Hz is the originate modem representation of logic level 1. Thus, the simple modem of Figure 5-6 has converted the RS-232-C definition of a logic 1 to the FSK definition of a logic 1. If the RS-232-C Transmit Data line goes to a level of +12 V, the top oscillator will be turned off and the bottom oscillator will be turned on. The bottom oscillator generates a frequency of 1070 Hz, which is the FSK definition of logic level 0. Thus, an RS-232-C logic 0 (+12 V) has been converted to an analog signal logic 0. Combining the outputs of the two oscillators with the gating function of the operational amplifier, the modem transmitter shown in Figure 5-6 produces an originate modem Transmit signal similar to that shown in Figure 5-5. This is an oversimplification of the Transmit circuits of a modem, but it will serve to illustrate the basic principles of modem transmitter design.

Modem   Receivers The   modem   transmitter   discussion   above addressed the modulation function of modems. Next we investigate the concepts underlying the demodulation function of a modem receiver. In particular, we will look at a simplified circuit diagram that shows how a modem converts into RS-232-C signals the analog frequencies that it receives from the telephone channel. Note that in the transmitter discussion we discussed the modulation function from the point of view of an originate modem and therefore used the originate modem transmit frequencies. If we are to maintain a constant point of view, namely, that of an originate modem, our current discussion should use the originate modem receive frequencies: 2025 Hz for receive logic level 0 or space, and 2225 Hz for receive logic level 1 or a mark condition.

Figure 5-7 diagrams a simplified modem receiver circuit. The incoming signal, which is composed of two different frequency components, is applied to the input of two filter networks. The function of the filter network is to separate the incoming signals into the individual frequency components. The top network, which is composed of two separate filters, filters out all incoming frequencies except for 2225 Hz and a narrow band of frequencies very close to 2225 Hz. This is accomplished by first passing the incoming signals past a band rejection filter.

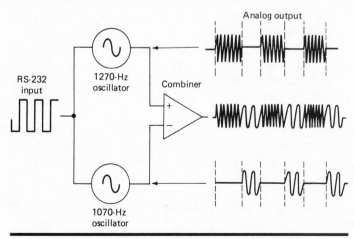

Fig. 5-6 Modem transmitter (modulator)—conceptual design.

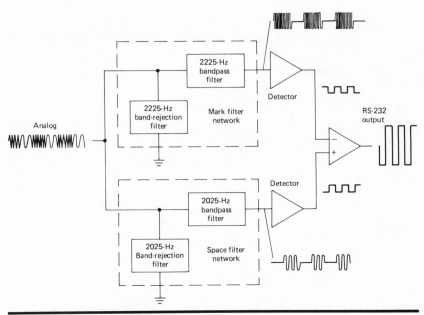

Fig. 5-7 Modem receiver (demodulator)—conceptual design.

The function of the rejection filter is to impose a high impedance (resistance to an ac signal) to the center frequency, in this case 2225 Hz. The output of this filter (sometimes called a *notch filter*) is tied directly to a ground. This means that any signal not in a narrow frequency range centered at 2225 Hz sees the rejection filter as a low-impedance path to a ground. The center frequency sees this filter as a high impedance to a ground and is relatively unaffected by it. The incoming signal passes next to the input of a bandpass filter. The function of the bandpass filter is to impose a high impedance on any frequency that is not close to its center frequency, in this case 2225 Hz, and to provide a relatively low impedance path for the center frequency. If the circuit is properly designed and tuned, the output will look (on an oscilloscope) something like the signal appearing in Figure 5-7 just to the right of the filter circuit. Ideally, this would be only the 2225-Hz portion of the incoming signal. In the real world, however, it is difficult actually to filter out all the unwanted components, but they can be adequately reduced by relatively simple filters made from operational amplifiers. The study of filters is a very complex subject and beyond the scope of this text. For the interested reader, any good college level text on reactance circuits will provide additional information; however, the study of filters will require some background in electronics.

The lower filter network in Figure 5-7 functions exactly the same as the one just described, except that the center frequency of the filters is 2025 Hz. Thus the output of this network, as shown to the right in the diagram, is the 2025-Hz portion of the incoming signal. This network then extracts the space (logic level 0) information from the FSK analog signal.

The outputs of the two filter networks are shown connected to a device called a *detector*. The function of the detector is to demodulate a signal. More specifically, in the circuit of Figure 5-7, the detectors will produce a positive output voltage when a sine wave is present at the input. When no sine wave is present at the input, the detector output goes to zero. Since we are modulating the analog carrier signal with a binary signal, the output of one of the detectors is always high. In fact, the output of one of the detectors is always the complement of the other. If the top detector has an active output, then the incoming analog signal is in a mark (logic 1) condition. If the bottom detector output is active, then the incoming analog signal is in a space (logic 0) condition.

The outputs of the two detectors are connected to the inputs of a simple operational amplifier. The function of the operational amplifier is to convert the two detector outputs to an RS-232-C signal suitable for transmission to a local device connected directly to the modem. This conversion is done by connecting the space detector output to the non-inverting input of the operational amplifier and connecting the mark detector output to the inverting input. This means that when the space detector output is active, the operational amplifier outputs a positive voltage. When the mark detector output is active, the output of the operational amplifier is a negative voltage. If we use a feedback resistor on the output of the operational amplifier, adjusted for enough gain to produce a 24-V swing (from $+12$ to -12 V), then we have instant RS-232-C! For those readers who are unfamiliar with the workings of an operational amplifier, a good reference is the *Radio Shack Engineers Notebook II*. This text gives basic information about operational amplifier operation for readers without extensive electronics backgrounds.

From the above discussion of modem transmitters and receivers, we may conclude that it is a little bit more difficult to extract digital information from an analog signal than it is to put it there. The modulation technique implemented in Figures 5-6 and 5-7 is called frequency shift keying (FSK) for an originate modem. The operation of an answer modem is functionally identical, with the only difference being the frequencies involved. Note that the Bell 103–compatible modems that are so common in microcomputer configurations use the FSK modulation technique.

Medium- and High-Speed Modems

Applications

The low-speed modems that we have discussed are normally used only for man-machine interfaces or for low-speed interfaces such as remote terminals and character printers. Low-speed modems, when used for computer-to-computer communications, quickly become the bottleneck of the system. For example, a low-speed modem that is transferring at 600 bps would take over 15 min to transmit a 56-kbyte data file, assuming that there are no errors and retransmission is not necessary. Since files of this size would not be considered large, it becomes apparent that for frequent file transfers a low-speed modem does not provide optimal support.

For groups of several terminals whose interactive traffic is concentrated and then communicated to a remote host via a cluster controller and a single modem, low-speed modems do not possess enough data transmission capacity to provide adequate support.

Typically, for computer-to-computer communications and for support of terminal clusters, medium- and high-speed modems are used as interfaces to the telephone network. In this section we discuss several modulation techniques that are most commonly used by these types of modems.

Modulation Techniques

The modulation technique used by a modem determines the maximum data rate, in bits per second, that the modem can support. For FSK modems, the maximum rate is 600 bps. By using more complicated modulation techniques, medium- and high-speed modems achieve higher data rates. Two of these basic modulating techniques should be familiar sounding since they are used by commercial radio stations for modulating radio frequency waves: *amplitude modulation* (AM) and *frequency modulation* (FM). AM and FM modulating techniques are not often used by themselves but are normally used by modems in conjunction with another modulating technique called *phase modulation*. However, before we discuss phase modulation, we will address the pure AM and FM techniques.

Amplitude Modulation (AM) An *amplitude-modulated analog wave* is a constant-frequency waveform in which the amplitude varies in step with the frequency of an impressed signal. This idea is illustrated in Figure 5-8. If we could see radio waves with no modulation imposed on the carrier, we would see a wave like that illustrated in Figure 5-8a. Figure 5-8b shows what an analog representation of some information

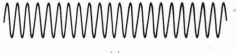

(a)

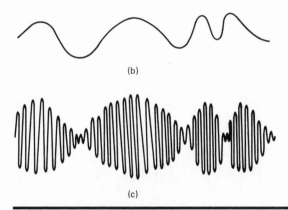

(b)

(c)

Fig. 5-8 Amplitude modulation concept: (a) unmodulated carrier, (b) analog modulating signal, (c) amplitude-modulated signal.

to be transferred might look like. If we take the algebraic sum of the amplitudes of the two waves, we have the resulting waveform shown in Figure 5-8c. This process is called *modulation by amplitude* or *amplitude modulation* (AM). To transfer digital data, the transmitter circuitry of an AM modem performs the conversion as described above, while the receiver circuitry recovers the modulation from the carrier wave. The circuitry necessary to perform the modulation and demodulation of the carrier wave is analog and does not usually rely on digital techniques. Most texts on electronic theory provide detailed explanations of AM circuitry.

Frequency Modulation (FM)   A *frequency-modulated wave* is a carrier wave whose frequency is varied by an amount proportionate to the amplitude of a modulating signal. Figure 5-9 illustrates this technique. Figure 5-9a shows an unmodulated FM carrier wave. Figure 5-9b shows an analog representation of some information to be FM-encoded. If the frequency of the carrier wave is varied in proportion to the amplitude of the analog modulating signal, the resulting waveform looks like the one illustrated in Figure 5-9c. As with AM, the modulation process for

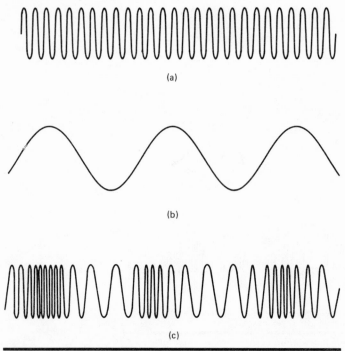

(a)

(b)

(c)

Fig. 5-9 Frequency modulation concept: (a) unmodulated carrier, (b) analog modulating signal, (c) frequency-modulated signal.

FM is performed by analog circuits, the discussion of which is beyond the scope of this text.

Phase Modulation Phase, or more precisely, phase angle, and its relationship to electrical signals constitutes the basis for phase modulation. *Phase angle* is defined as the difference in degrees between corresponding stages of progress of two cyclic operations. Let us again look at an illustration.

Figure 5-10a shows one full cycle of a single sine wave. This full cycle can be divided into 360 increments called *degrees*. The sine wave starts at 0 degrees (0°) with an amplitude of 0. Its maximum positive amplitude is reached one-fourth of the way through the period at 90° (360° ÷ 4 = ʼ90°). The maximum negative amplitude is reached at 270°, or three-fourths of the way through the period. In talking about signals, it is more convenient, however, to refer to the phase angle, or to a particular point on a wave, than to positive or negative amplitude. Therefore, if we refer to a sine wave phase angle of 180°, we mean the point halfway through the period with an amplitude of 0.

We can now use this technique to compare two waveforms at the same point in time by referring to the difference in phase angle between the waves at that point in time. Referring to Figure 5-10b, we see two waves that have a phase angle difference. At time t_1 the lower wave has a phase angle of 90°, while the upper wave has an angle of 0°. At time t_3 the lower wave has an angle of 270°, while the upper wave has an angle of 180°. At these two times and at all other times along the time scale, the difference between the phase angles of the two waves is 90°. We can therefore say that the phase angle difference between the two waves is 90°.

PHASE SHIFT KEYING (PSK): In discussing the concept of phase angle shift in relation to a single waveform, we can make use of the technique just covered. If we look at Figure 5-10c, we see a waveform that has been

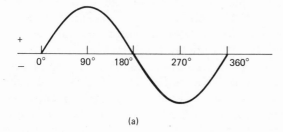

(a)

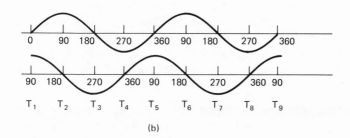

(b)

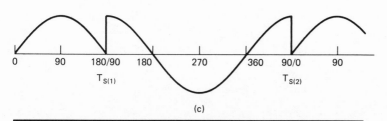

(c)

Fig. 5-10   Signal phase concept: (a) one period of a sine wave, (b) two sine waves with a phase angle difference of 90°, (c) single signal shifted 90°.

shifted $90°$ at the point labeled $T_{S(1)}$. At that point the wave has been shifted back $90°$ from an angle of $180°$ to an angle of $90°$. This same $90°$ shift occurred at the point labeled $T_{S(2)}$. Here the wave was shifted from $90°$ back to $0°$. Thus, for one full period, namely, the $360°$ between $T_{S(1)}$ and $T_{S(2)}$, the waveform has been perturbed by a phase angle of $-90°$.

If we had the capability to shift an electrical sine wave by four different phase angles and then detect those angles at some remote location, we could increase the throughput of a modem-driven communication line without greatly increasing the frequency of the electrical signals. This is a technique used by PSK-type modems. Original PSK-type modems, such as the Bell 201B, shifted the signal by $45°$, $135°$, $225°$, and $315°$. These phase shift angles were relative to the previous cycle and not to the original phase angle of the signal. Each of the four different phase angles was used to represent 2 bits instead of 1 single bit as in FSK modulation. The four angles represent the four possible combinations of 2 bits: 00, 01, 10, and 11. These bit combinations are referred to as *dibits.* The use of this technique has the effect of doubling the line transmission rate; that is, if the modem transfers 600 dibits per second, the effective throughput rate is 1200 bps. Newer modems, like the Bell 201C, are capable of shifting a signal by as many as eight different phase angles, representing all the different combinations of 3 bits. By coupling this capability with a line signal rate of 1600 baud, these new modems have an effective throughput rate of 4800 bps. Notice the difference between bit rate and baud, or line signaling, rate for the various modulation techniques. Table 5-1 illustrates the differences. The numbers in parentheses in the table represent sample signaling (baud) rates and associated bit rates for the various modulation techniques.

PHASE AMPLITUDE MODULATION (PAM): Due to the characteristics of the transmission media, the number of multiple phase shifts (2, 4, 8, etc.) that can be used to encode data quickly reaches an upper limit. This

TABLE 5-1

Modulation technique	Line signaling rate, bauds	Bit rate, bps
FSK	n (300)	n (300)
Dibits	n (1200)	$2n$ (2400)
Tribits	n (1200)	$8n$ (9600)
Quadbits	n (1200)	$16n$ (19,200)

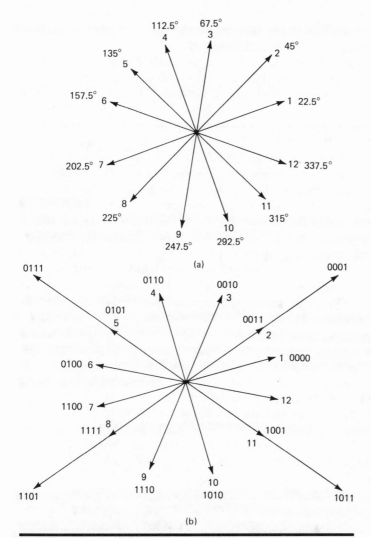

(a)

(b)

Fig. 5-11 Phase amplitude modulation (PAM): (a) twelve phases used for PAM, (b) PAM quadbit encoding.

problem can be partly overcome by combining the PSK and AM modulation techniques. Current, commercially available modems are capable of representing the 16 discrete combinations of 4 data bits. Modems of this type can achieve effective data rates of 9600 bps. Figure 5-11a is a phase angle diagram showing the 12 phases used for phase amplitude modulation (PAM). If we add two amplitude levels to four of these angles, we have 16 discrete angle or amplitude levels. Figure 5-11b shows the 16 phase angle levels and the 16 bit patterns they represent.

High-performance modems that employ PAM techniques are normally used for synchronous transfer applications.

This concludes our characterization of modem types and classes according to speed and modulation technique. Some additional ways in which to classify modems are briefly described below.

Two- and Four-Wire Modems

A third way to classify modems (in addition to speed and modulation technique) is as a two- or four-wire operation. The choice between two- and four-wire operation is determined by the need for full- or half-duplex operation, with one exception—low-speed modems. The narrow bandwidth used by low-speed modems enables them to achieve full-duplex communications within the 3000-Hz band of the public-switched telephone network. However, the medium- and high-speed modems that use more complex modulation techniques require the full 3000-Hz public telephone network bandwidth to communicate in just one direction. Therefore, most applications that require high-speed data transfers in two directions concurrently use four-wire modems.

High-performance two-wire modems implement a medium- or high-speed primary channel that supports transmission in one direction plus a very low speed secondary channel that transfers in the opposite direction. The function of the secondary channel is to provide rudimentary handshaking. In particular, the secondary channel carries signals that acknowledge the receipt of data blocks on the primary channel. The secondary channel is often called the *reverse channel.* This type of modem is useful in applications for which the main flow of data is in one direction and the reverse channel is used only for control information.

In applications requiring medium- or high-speed data flow in both directions concurrently, a four-wire circuit is usually required. The use of four-wire circuits usually, but not always, involves the installation of local private lines. Four-wire circuits can be dedicated or switched. In applications that are susceptible to noise, the dedicated circuits offer better noise immunity. Four-wire dedicated circuits can be provided with signal conditioning to reduce noise further. In most applications that use unconditioned switched lines, the cost difference between two- and four-wire circuits is so small that four-wire circuits often turn out to be more cost-effective.

Short-Haul Modems

All of the modems that we have discussed so far are meant to interface with the public-switched telephone network. Another type of modem is the short-haul modem. Short-haul modems are not designed to be interfaced through any common carrier or switched network. Modems in this class are directly connected through cables. This direct wire connection, however, limits the distance that can separate the originate and answer short-haul modems. This distance is usually limited to something less than 10 mi. Modems in this class are sometimes referred to as point-to-point modems. When communicating over very short distances, these modems can reach bit rates of 1 Mbps. The transfer rate of short-haul modems is inversely proportional to the length of the connecting cable.

Short-haul modems use many different techniques for transferring information. Some use standard modem modulation methods, while others transfer the digital signals directly without converting them to analog signals. They often use balanced lines, or differential amplifiers, to do this. Since short-haul modems are normally purchased in pairs and are almost never attached to a shared network, the manufacturers are free to use whatever transfer technique is desired.

Short-haul modems are often found in industrial plant process-control systems. Computer manufacturers also use short-haul modems for connecting central processors together in a local network. Here the distance between nodes is usually less than 100 ft, and the modems can operate at full speed. Due to the diverse nature of short-haul modems, any application of them should be thoroughly investigated with the various vendors of these devices.

Bus-Compatible Modem Units

The final type of modem that we will describe is the single-board bus-compatible unit. While modems of this type usually fall into one of the other categories that we have already discussed, their unique feature of bus compatibility deserves special mention. These single-board modems are becoming very popular in the microcomputer area. This may be due in part to standardization of system buses, which is not found in other types of computers. Single-board modem units also take advantage of state-of-the-art technology to cost-effectively provide many useful capabilities. A typical single-board modem that plugs into the backplane of an S100-bus microcomputer system provides autoanswer, autodial, originate and answer modes, with direct connection to the telephone network. Although most modems in this class subscribe to the Bell 103 mode of operation, some are available that transfer at 1200 baud. It is not inconceivable that modems in this class may someday be included as part of a standard serial I/O board and will transfer at ever higher data rates. Indeed, modems are now being manufactured that are an integral part of a standard telephone set (such as the Racal-

Vadic Modemphone). In fact, modem functions may soon be included as a programmable option of a UART.

MODEM CONTROL SIGNALS

The above discussion has addressed the basic techniques that modems use to modulate and demodulate signals that they send and receive over the telephone network. In addition to performing this conversion function, modems must also control the operation of the interface between the local equipment, such as computers and terminals, and the communication line. In some cases this can be a trivial task, while in others the control functions are reasonably complex and require specialized circuitry to implement.

For example, *autodial* modems have the ability to dial phone numbers of remote systems and establish a connection without manual intervention. An *autoanswer* modem has the ability to answer an incoming call on the phone line to which it is connected and to open a channel to the locally attached computer. Autodial and autoanswer modems must have the intelligence to control both the phone lines and the attached equipment. In the case of the autoanswer modem, the phone connection cannot be established if the local system is not ready to transfer data. This modem must be able to sense the status of the local system and make a decision as to whether or not to answer the incoming call. In the case of the autodial modem, the modem must initiate the phone call and detect an operational system on the other end before enabling the local system to begin transferring data.

Many modems have built in diagnostic capabilities. These diagnostics not only check out the modem but also allow testing of the communication line and the channel to the local system. All of these capabilities require the use of the RS-232 control signals. The circuitry required to implement the RS-232 control functions described in Chapter 3 can be quite complex. Consistent with the trend to implement specialized integrated circuits (ICs) to perform various communication functions, there are now commercially available IC components tailored for modems.

SPECIALIZED MODEM CIRCUITS

Motorola Semiconductor Products Inc. offers an integrated circuit CMOS (complementary metal-oxide semiconductor) device that contains most of the circuitry necessary to construct a low-speed 103-type modem. The device, a 16-pin dual-inline package, is called an MC14412

Universal Low-Speed Modem. Making a fully operational modem based upon this chip requires the addition of a bandpass filter, a power supply, and a crystal. The chip sells for less than $10 (1981 prices) in large quantities. For those readers who are interested, information about this chip can be obtained from Motorola Semiconductor Products Inc., Box 20912, Phoenix, Ariz. 85036.

Bandpass filters for low-speed modems can also be purchased off the shelf for connection to devices like the MC14412. Filters, however, are expensive when compared to the rest of the circuits required for modems. Prices of $50 (1981) for the required filters in large quantities are not uncommon.

An important fact to consider concerning the decision about whether to build your own modem or purchase a commercial modem is that FCC regulations require that the unit be tested and approved prior to connection to the telephone network. The modem approval process is a long and costly effort that is cost-effective only for manufacturers of large numbers of modems, for whom the cost incurred can be amortized over many units.

Consideration of Federal Communications Commission (FCC) jurisdiction and regulations over the uses that may be made of microcomputer systems brings us full circle back to why we need modems in the first place—to be able to utilize the public-switched telephone network for data communications. The common carriers (most notably AT&T and the Bell System Operating Telephone Companies) have invested many years and dollars to create this incredible resource. A number of landmark decisions in the courts and in the development of FCC policy have given individuals and companies a mandate to lease telephone resources for data transmission. In consideration for being able to use this resource, the FCC and the common carriers require that the user do nothing to harm the system physically or to impair operations of others who share this resource. Thus, the requirement for approval of modems prior to use (called "type acceptance") is designed to be a control over what electronic equipment can be tied to the telephone system.

Clearly, the telephone system is a major factor in the types of data communications applications that are the subject of this book. The next section discusses the characteristics of this system that are especially important for data communications applications.

THE PUBLIC TELEPHONE NETWORK

The most important characteristics of the public-switched telephone network for a user of data modem equipment are the traffic capacity, circuit interface characteristics, noise, and bit error rate statistics.

Capacity

It is easy to get the impression that the phone system can handle an unlimited number of calls. In making telephone calls for voice communication, we always obtain a dial tone when we pick up the receiver and are able to dial the number of the other phone. The instance of a busy signal is relatively rare and most often occurs because the person that we are trying to reach is already using the phone. By design, it is not at all obvious that most telephone networks can handle only about 20 percent of the subscribers at any one time. In fact, some rural networks may have the capacity to handle only about 10 to 15 percent of the subscribers. Since most phones are only used for a small percentage of the time, it is not efficient to provide the capacity to handle 100 percent of the possible traffic all the time. None of us would care to pay the cost of the equipment required to provide this kind of service.

Connectivity

In 1979 the American Telephone and Telegraph Company (AT&T), sometimes referred to as "Ma Bell," sometimes referred to as the telephone company, and occasionally referred to by other names, reported over 140 million telephone sets in service across the United States. If it were necessary to string one pair of wires for every two telephones in existence, then the number of wire pairs that would be required is approximately 9800×10^{12}. That is almost 10 million billion wire pairs. Clearly, this would not be feasible. In fact, AT&T provides the same *connectivity* with only 80.6 million lines. Since precisely the same connectivity is provided with only a fraction of the number of lines that would be required if dedicated wire pairs were used between each pair of telephones, there must be a great deal of resource sharing going on.

Resource Sharing

The telephone network is an example of a shared resource. The sharing occurs at two levels, on a *time division basis* and on a *frequency division basis*. *Time division sharing* occurs on each telephone call and is called *circuit switching*. This is diagrammed in Figure 5-12. A real circuit is set up for a telephone call between station A and station B. To set up such a call, the switches at each of the intervening switching centers (labeled 1, 2, and 4 in Figure 5-12) must be set to the proper position to provide a continuous circuit connection from station A to station B for the complete duration of the phone call. The circuit is set up once and, with the switches in place, the circuit "belongs" to the station pair (A and B) for the duration in time of the telephone call. The circuit is taken down once the phone call is over.

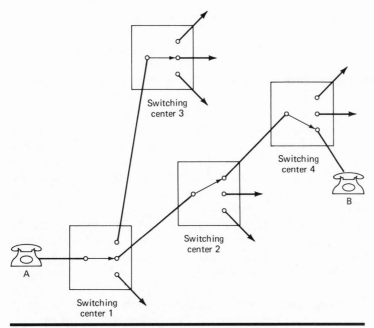

Fig. 5-12 Circuit switching in the public telephone network.

A second level of sharing occurs on a *frequency division basis*. That is, each of the wires between the switching centers in Figure 5-12 carries many telephone calls *simultaneously*. Each telephone call is placed on a separate frequency, which is multiplexed with many others of different frequencies, so that they do not interfere with each other when they are transferred over the same wire. This requires relatively sophisticated equipment at the switching centers but provides an additional level of sharing for the switched telephone network. Telephone company terminology describes the level of frequency division multiplexing in terms of *groups, supergroups,* and *mastergroups:* a *group* multiplexer supports 12 individual voice channels on a single 48-kHz channel; a *supergroup* consists of five groups, or 60 voice channels; and a *mastergroup* consists of 10 supergroups, or 600 voice channels. The equipment required to implement groups, supergroups, and mastergroups is sophisticated and expensive; however, it provides an additional level of sharing for the telephone network resources.

Bandwidth
Bandwidth is a critical characteristic of the telephone network that influences capacity. *Bandwidth* is defined as the difference between the highest and the lowest frequency used for transmission. The larger the

bandwidth of a system, the greater data throughput the system will support.

The public-switched telephone system was designed for use as a human voice transmission network and therefore has a frequency range of 300 to 3300 Hz, which covers the lower portion of the range of normal human speech. Actually, the frequency components of human speech range between 20 and 20,000 Hz. The human ear responds to frequencies from 30 to 30,000 Hz; however, a very high percentage of the energy content of human speech occurs in the range of 30 to 3500 Hz, as is shown in Figure 5-13, a plot of frequency versus energy in human speech. Note that the interval between 300 and 3300 Hz on the horizontal axis clearly contains the range in which the greatest energy values appear on the vertical axis.

The range of 300 to 3300 Hz (shown in Figure 5-14) represents a compromise that is reasonably economical yet provides an adequate bandwidth to make human voice communication over the public telephone network possible. A communication channel with a larger bandwidth would provide more "information"-carrying capacity, since more of the frequency components of speech could be transferred. However, equipment that will reproduce and respond to a larger frequency range is more expensive. Thus there is a trade-off between expense and information-carrying capacity. For standard voice channels, the compromise that best balances expense with capacity is the 300- to 3300-Hz range. Knowing the frequency band supported by the telephone network, we may compute the bandwidth to be 3300 Hz − 300 Hz = 3000 Hz.

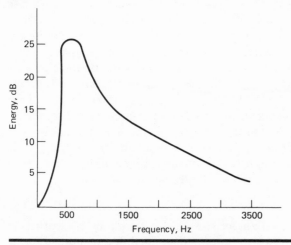

Fig. 5-13 Energy distribution in human speech.

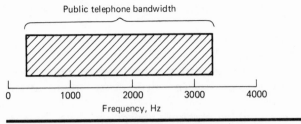

Fig. 5-14 Bandwidth of the public telephone network.

Interface Characteristics

A major area of interest to modem users is the interface characteristics of the phone system. There are two methods of connecting modems to telephone lines: the first method is to use acoustic coupling, and the second is to use direct coupling. Acoustic-coupled modems require no special interface to the public telephone network, while directly coupled modems require some direct-access arrangement (DAA) interface.

FCC Regulations

As a user or designer of modem-based communication systems, you must be aware that all direct connections to the public telephone network must be approved by the Federal Communications Commission (FCC). In addition to direct connections, the levels of signals induced into the public telephone network by acoustic-coupled modems are controlled by the FCC. All modems approved for use on the public network will bear an FCC-type acceptance number and should be the only modems considered for use.

Acoustic Coupling

Acoustic coupling is implemented on the transmitter side of a modem by driving a standard telephone handset transmitter with a speaker or transducer. The receiver side is implemented by picking up the telephone handset receiver signals with a microphone. This is usually done by mounting the speaker and microphone in one end of two cylindrical rubber grommets. These two grommets are then mounted so that a telephone handset will fit into the two open ends of the grommets. This forms an acoustically tight coupling between the handset and the modem pickup and driver.

This type of modem connection can be implemented on any available standard telephone set. Thus, an acoustic modem can be used anywhere there is a telephone. This is a favorite coupling technique for portable terminals with built-in modems. Acoustic coupling is limited in the bandwidth and noise immunity it can provide and is therefore used only in low-speed applications.

One application of a portable terminal with built-in modem might be for an insurance salesperson to bring the terminal into a client's home and establish a connection with a computer in the main office. This would allow the salesperson to provide instant information on various policy calculations.

In the past the acoustic-coupled modem allowed access to the telephone network without the cost of leasing an interface from the telephone company. However, recent FCC rulings now allow direct connection to the network with FCC-approved devices without having to lease a direct-access arrangement (DAA) from the telephone company. Thus, direct-connection modems can compete in cost with acoustically coupled modems and are becoming increasingly popular in low-speed applications. In particular, many FCC-approved direct-connection low-speed modems have been developed for the microcomputer systems market.

Direct Coupling

Direct coupling to the public telephone network is accomplished with a DAA device. This device is usually (but not always) provided by the local telephone company and must be leased on a monthly basis. The type of DAA device needed depends on the type of modem used. Autoanswer and autodial modems require different interfaces than manually controlled modems. The local telephone company data-services representative is an excellent source of information concerning the types of DAAs that should be used.

Echo Suppressors

To reduce the unwanted effect of echoes on the voice-grade circuits of the long-distance network, the common carriers have installed electronic echo suppressors. Electronic echo suppressors allow signals to flow only in one direction at a time. When someone is talking (or signaling) on the circuit, the suppressor stops the flow of signals in the opposite direction. When the first talker stops talking, the suppressor senses this and turns the line around, thus allowing the second person (on the other end) to talk while suppressing echoes flowing back to the second talker. The time required for the echo suppressors to turn the line around may be on the order of 300 ms. While this may go unnoticed for voice communications, such a delay becomes significant for half-duplex data communications. Note that, at 300 bps, 300 ms is enough time to transmit nine complete characters.

Full-duplex communication through an active echo suppressor is impossible. The echo suppressor can be disabled, however, by applying a 2100-Hz tone to the circuit for 400 ms. The suppressor will remain disabled as long as signals are present on the circuit. If the data signals are lost for a period exceeding 100 ms, then the echo suppressor is again enabled. One common way to avoid the problems that echo suppressors can cause is to implement the communications circuit as a four-wire channel.

Distortion and Noise

In the context of this book, the medium for long-distance communications is the telephone network. Clearly, a minimum requirement for a medium is that it accurately convey messages from a source to a target destination. Unfortunately, very few media are totally error-free. So the question is not whether or not the medium is perfect, but, first, why isn't it and, second, how close is it to perfection. The following paragraphs address these questions for the telephone network.

Unit of Measurement for Distortion: The Decibel

In dealing with communications, the term *decibels* is often encountered in discussions of signal levels in transmission circuits. The decibel is actually the ratio of the output of a circuit to its input, expressed as $10 \log_{10}$ of that ratio. This is simply a method of expressing the gain or loss of that circuit. For example, if we apply a signal level of 0.1 W to a circuit and we observe an output of 1 W, then we can say the circuit has a gain of 10 decibels (dB):

$$10 \log_{10} 1/0.1 = 10$$

If the output of this circuit were 0.2 W, then the gain expressed in decibels would be approximately 3 dB:

$$10 \log_{10} 0.2/0.1 = 3.01 \text{ dB}$$

Had the output of the circuit been 0.01 W, with an input of 0.1 W, the gain would have been -10 dB:

$$10 \log_{10} 0.01/0.1 = -10 \text{ dB}$$

In this case, we would say that the circuit has a loss of 10 dB.

The important thing to remember is that measurements expressed in decibels are relative. For example, if we have a communication line that has a loss of 10 dB, we must know the input power level before we can calculate the output level. If we apply an input power level of 2 W to this circuit, we can expect an output power level of 0.2 W. A general rule of thumb is that *power* doubles (approximately) every +3 dB and halves (approximately) every −3 dB. If we are expressing the gain or loss as a function of *voltage* or *current*, then the equation for decibels becomes 20 $\log_{10}$ of the ratio of input to output voltage or current. Therefore, voltage and current double or halve every 6 dB.

Attenuation Distortion and Envelope Delay Distortion

Signals transmitted over the public telephone network will experience some degree of distortion. A signal that is transferred over a voice channel is said to be *distorted* if the output signal from the channel is not a precise replica of the channel input signal. Signals that are transmitted over the public telephone network are susceptible to two different types of distortion. The first of these is *attenuation* or *amplitude distortion.* Attenuation distortion is the gain or loss in amplitude of a signal that is not uniform across all frequencies in the frequency range or bandpass. The second type of distortion occurs because propagation delays for signals from source to destination are not uniform across all frequencies in the spectrum. This type of distortion is called *envelope delay distortion.*

Amplitude distortion and envelope delay distortion occur together in the telephone network when microwave (superhigh-frequency radio) equipment is used. Since microwave energy travels (more or less) in straight lines and since the earth is round, the typical AT&T Long Lines configuration has a microwave repeater station every 25 mi, on the average, along its long-distance communication path. These stations accept incoming microwave signals and redirect them toward the next

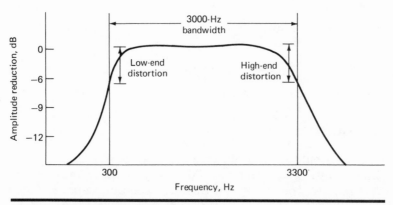

Fig. 5-15   Attenuation distortion.

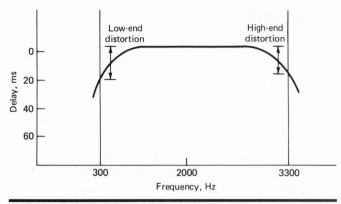

Fig. 5-16 Envelope delay.

repeater station along the path. Thus, almost any time that data transfers are made over long distances, the transfers involve the use of microwave repeaters. Such devices tend to reduce the bandwidth of the transmission channel slightly because of filtering by the amplifiers of the microwave equipment. In some cases the usable frequency range may be between 500 and 2900 Hz only. This bandwidth limitation is the result of signal amplitude distortion. The circuits used in the microwave equipment attenuate, or reduce the amplitude of, the signals on the high and low ends of the usable frequency spectrum. Figure 5-15 shows the effect this amplitude distortion has on the spectrum of the public telephone network.

In addition to amplitude distortion, these circuits also induce an envelope delay, again at the high and low ends of the usable spectrum. This envelope delay is the actual propagation delay of these outer frequencies. Figure 5-16 shows the extent of the envelope delay distortion imposed on the public telephone network. Both the amplitude distortion and the envelope delay cause the bandwidth to be reduced.

Both amplitude distortion and envelope delay distortion problems can be overcome by employing a technique called *equalization.* For amplitude distortion problems, equalization is accomplished by providing extra amplification of the outer frequencies prior to transmission over the phone lines. If the proper level of amplification is used, the received signals will be of equal amplitude. Equalization can also be accomplished by reducing the amplitude of the frequencies near the center of the bandwidth prior to transmission of the signals. To overcome envelope delay distortion problems, the equalization technique delays transmission of the center frequencies, so that, to the receiver, all received signals seem to have the same propagation delay.

The distortion just described can also affect leased private lines; however, the common carriers provide various levels of line conditioning to diminish these problems. Conditioning is achieved by using special amplifiers that match transmission levels and impedances. Obviously, the more signal conditioning an application requires, the higher the lease rates of the line will be.

Random Noise and Impulse Noise

Noise is another big problem with data communications across the public telephone network. There are two major classes of noise: *random* and *impulse*. *Random noise* is normally associated with line length; it increases as the length of the communication line increases. *Impulse noise* is caused by the line-switching equipment used by local phone companies. Noise problems are difficult, if not impossible, to solve. Therefore, applications that require highly reliable communications use leased conditioned lines to eliminate as many of the usual noise sources as possible.

Other Performance Factors

Other factors affecting the performance of modems are *phase jitter, harmonic distortion,* and *crosstalk. Phase jitter* is the continuous shifting of the phase or frequency of the individual frequency components of a signal. *Harmonic distortion* is a condition that exists in the output of an amplifying circuit when harmonics, which are added during the process, alter the transmitted signal waveform. Most modems are not seriously affected by these problems, however. *Crosstalk* is interference with a transmitted signal that is caused by stray electromagnetic coupling from nearby circuits. The amount of crosstalk interference is dependent upon the type of equipment used by the local phone company. As the use of electronic switching and fiber optics increases, the amount of this type of interference will diminish. Currently, these problems are solved by using sophisticated modulation techniques.

Bit Error Rate Statistics

Errors on a communication channel are caused by a variety of different factors. The error rate of a given channel will depend on the type and speed of the modem being used, the type of lines being used (switched or conditioned), and the data path taken through the public telephone network. So many factors have an effect on the communication channel that it is difficult to predict the probability of errors occurring during a specific transmission. For example, a transfer using low-speed modems over long-distance switched lines will experience different error rates if the call is routed through different exchanges.

In 1969 and 1970, the Bell System conducted an extensive survey in an attempt to characterize the switched telephone network in terms of character errors and lost characters. The part of the survey that dealt with low-speed transmission involved a configuration that transmitted a message over telephone circuits using a Bell 103 modem operating at 150 bps. The survey emphasized character errors, as opposed to bit errors, because most low-speed terminals and character printers are character-oriented. They send and receive data on a character-by-character basis and are usually connected to computer-based applications that operate on characters, rather than bits. In addition, owing to the nature of the asynchronous (start/stop) protocol, a single erroneous bit can potentially cause several character errors.

The Bell study set up several dozens of communicating station pairs, sent messages between them, compared the messages sent with the messages received for over 500 nominal 40-min calls, and statistically analyzed the data. The sending and receiving station pairs were categorized into "mileage bands" according to the distance that separated them: the short mileage band consisted of pairs within 180 mi of each other; the medium mileage band covered distances between 180 and 725 mi; and the long mileage band consisted of station pairs separated by 725 to 2900 mi. The U.S. stations participating in the survey were located in the 48 contiguous states, and the Canadian stations, in the provinces of Ontario and Quebec.

Average Error Rates

Table 5-2 summarizes the results of the Bell System survey for average character error rate and average character loss rate. The *average character error rate* is defined as the ratio of the number of character errors to the number of characters sent. Thus, the character error rate of 1.46×10^{-4}, given as the average across all calls, implies that in 10,000

TABLE 5-2 *Bell survey results for average character error rate and average character loss rate*

	All calls	Mileage band		
		Short	Medium	Long
Total number of calls	534	171	186	177
Characters transmitted ($\times 10^3$)	21,310	7,000	7,480	6,830
Character errors	3,110	751	1,053	1,306
Lost characters	14,511	9,581	1,476	3,454
Character error rate (per 10,000)	1.46	1.07	1.42	1.90
Lost character rate (per 10,000)	6.81	13.7	1.98	5.03

characters transmitted an average of 1.46 character errors occurred. This corresponds to one erroneous character per every 1.7 pages of single-spaced typed text.

The *lost character rate* (or *character loss rate*) is defined as the ratio of the number of characters for which no bits were received (usually caused by a loss of the received signal) to the total number of characters transmitted. The average lost character rate for all calls was 6.81×10^{-4}, or 6.81 characters lost for every 10,000 sent. This corresponds to 2.7 missing characters per page of single-spaced typed text.

The columns labeled short, medium, and long give the survey statistics for the three mileage bands described above. Note that the character error rate is poorest (highest) for the long-distance transmissions, while the lost character rate is poorest for the short-distance transmissions.

Error Burst Properties

Transmission errors are inherently bursty. By this we mean that they tend to occur in groups. More precisely, given that an error occurs, the

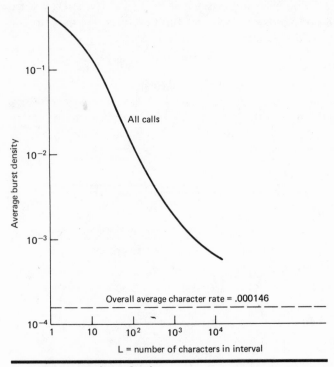

Fig. 5-17 Average burst density.

probability of additional errors occurring in the next several characters is much higher than the overall average character error rate.

To analyze the degree to which the errors that were documented in their 1969–1970 survey tended to occur in bursts, Bell System analysts defined a quantity called *burst density*, which is defined as the error rate over an interval of time associated with transmitting n characters. As n increases, the burst density approaches the overall average character error rate, which is based upon the total number of characters sent. Figure 5-17 shows a graph that dramatically illustrates the difference between burst density and overall average character error rate. What the graph shows is that for small values of n, say, 10 characters or less, the occurrence of one error implies that the likelihood is quite high that there will be an error in the next 10 characters to be transmitted.

The same Bell survey also investigated the transmission errors associated with medium- and high-speed transmission. The average character error rate obtained for a Bell 103 modem operating at 1200 bps was observed to be 6.6×10^{-4}.

The character error rates computed and reported in the Bell survey can be converted to bit error rates if we assume that there is exactly 1 bit error associated with each character error, and that each character is composed of 10 bits. Under these assumptions, we can obtain the bit error rate from the character error rate by subtracting 1 from the exponent on 10. For example, an average character error rate of 1.46×10^{-4} converts to an average bit error rate of 1.46×10^{-5}.

For more detailed information concerning how the Bell System survey was performed and its results, we refer you to the two major articles that published its findings:

H. C. Fleming and R. M. Hutchinson, Jr., "1969–1970 Connection Survey: Low-Speed Data Transmission Performance on the Switched Telecommunications Network," *The Bell System Technical Journal*, vol. 50, no. 4, April 1971, pp. 1385–1406.

H. W. Klancer, S. W. Klare, and W. G. McGruther, "1969–1970 Connection Survey: High-Speed Voiceband Data Transmission Performance on the Switched Telecommunications Network," *The Bell System Technical Journal*, vol. 50, no. 4, April 1971, pp. 1349–1384.

While the telephone network, modem technology, and computer component technology have undergone tremendous changes since these articles were written, they still provide good insight into what can be expected with respect to error rates in transmitting data via the telephone network.

ABOUT THE AUTHORS

DR. ELIZABETH A. NICHOLS *is president of Digital Analysis Corporation and specializes in the design and development of real-time systems for applications in data communications and computerized industrial process control.*

DR. JOSEPH C. NICHOLS *is vice-president of Digital Analysis Corporation and is responsible for the analysis, design, and implementation of computer communications systems.*

Elizabeth and Joseph have coauthored four books on microprocessor programming/interfacing and data communications. They present seminars on these subjects in Europe and the U.S.

KEITH R. MUSSON *is a senior engineer for ConTel Information Systems and specializes in microcomputer design and applications; hardware selection, evaluation, site preparation, and installation; and real-time systems development. Mr. Musson has coauthored two books on data communications for microcomputers and microprocessor programming and interfacing.*

ABOUT THIS BOOK

The manuscript for this book was prepared on a Cromenco microcomputer, using the Screen Editor utility available under the CDOS operating system. The output disks were converted to nine-track magnetic tapes, which were used to drive an APS-5 CRT typesetter via a Penta front-end composition system.

Index